COMPETITION AND
THE STRUCTURE OF
Bird
Communities

MONOGRAPHS IN POPULATION BIOLOGY

EDITED BY ROBERT M. MAY

COMPETITION AND

THE STRUCTURE OF

Bird
Communities

MARTIN L. CODY

PRINCETON, NEW JERSEY
PRINCETON UNIVERSITY PRESS
1974

Library of Congress Cataloging in Publication data will be found
on the last printed page of this book

This book has been composed in Monotype Baskerville
Printed in the United States of America
by Princeton University Press, Princeton, New Jersey

TO ROBERT MACARTHUR

Acknowledgments

In the course of writing this book, and particularly in conducting the field work it describes, I have become indebted to a great many people and take this opportunity to thank them. Pristine habitats for bird studies are becoming increasingly hard to find on our shrinking Earth, but this search has been facilitated by many cooperative landowners and stewards; in particular the United States National Parks and National Forest Services have been most helpful, and work in the Santa Monica Mountains was possible through the kindness of Miss Kathleen Murphy. I have been provided generous facilities at the Southwestern Research Station in Arizona, by Floyd Clarke at the Jackson Hole Research Station, by Robert Engel in Colorado, by Braulio Araya and Patricio Sanchez in Chile, and by the Regents of the University of California.

A number of colleagues have given me permission to use their unpublished or published material, and for this I am grateful to James Brown, Christopher Cody, Peter Grant, Kimball Harper, Henry Hespenheide, Jon Roughgarden, Hartmut Walter, and Richard Yeaton. The following publishers have given permission to reproduce figures and tables: Academic Press, Allen Press, Duke University Press, *Science,* and the University of Chicago Press. I have greatly benefited from the advice of a wider group of colleagues, who have mentioned in correspondence or conversation examples or counterexamples of various phenomena I discuss in the monograph, or who have offered additions and corrections to earlier drafts of its manuscript. For such invaluable advice I thank Steve Arnold, Ruth Bernstein, James Brown, Robert Cheke, Kenneth Asplund, Jared Diamond, J. Blondel, John Ebersole, Peter Grant, Richard Hutto, Robert Jenkins, Robert May, Philip Regal, Dennis Paulson, Henry Thompson, and François Vuilleumier.

ACKNOWLEDGMENTS

I am happy to acknowledge the assistance of artist Karl Pogany, who prepared most of the figures, including all of the more artistic ones. John and Barbara Pleasants helped to prepare the manuscript, and the UCLA secretarial staff under the direction of Barbara Munn provided expert typing service. My field work has been supported by a number of organizations, to which I express my sincere thanks: the Ford Foundation through the University of California-University of Chile Convenio, the Frank M. Chapman Fund of the American Museum of Natural History, the Jessop Fund of the Philadelphia Academy of Natural Sciences, the National Science Foundation, the New York Zoological Society, the Society of the Sigma Xi, and the Research Council of the University of California. I am also most grateful to the less easily quantified assistance from my wife Jane.

I mention finally the special debt I owe to Robert MacArthur, whose wonderful stimulation has been the source of great personal and professional satisfaction to me.

Contents

COMPETITION AND
THE STRUCTURE OF
Bird
Communities

CHAPTER ONE

Resource Division and the Niche

I. INTRODUCTION

A. *Different Sorts of Field Ecology*

The goal of ecology is to provide explanations that account for the occurrence of natural patterns as products of natural selection. Variation at the level of the phenotype, population, or community must be related to variation in some aspect of environment; two approaches have been employed, that of experimentally induced environmental variation and the use of variation which occurs naturally in the environment. Although both methods achieve results, the latter, in which the ecologist gathers observational and mensural data that either discredit or support the explanations, has been obviously successful. Although lacking the definitive aspects of cause-and-effect demonstrations under controlled conditions from which experimental and manipulative studies benefit, the approach has the advantages of a rapid progression through hypothesis and theory to acceptable and accepted explanations for natural phenomena with economy of time and effort.

Two further considerations have been given momentum to the acceptance of this sort of science: a great many questions and problems simply are not amenable to an experimental approach—many ecologically interesting variables such as climate, competitive milieu, foliage structure, and food level just do not lend themselves to manipulation. Even when such factors are amenable to alteration, there are severe difficulties in arranging for effective controls in field situations.

The obvious way around these difficulties in the experimental approach, and that usually adopted, is to make use

3

of the variations in environmental factors which occur naturally—the so-called "natural experiment." Such variation does not differ conceptually from that induced by an experimenter, and may in fact be a more appropriate test in some instances. Further, whether change is induced artifically or occurs naturally, the process by which scientific progress is assured—the progression of hypothesis, to test, to new or extended hypothesis—remains the same, and indeed is common to all fields of science.

It is apparent at this stage of human evolution that ecological science is being urged to produce workable relationships which can aid our own disastrous environmental predicament; ecology has been put on the spot whether we like it or not. Thus any and all insights are at a premium; the detailed infrastructure of broad relationships can be illuminated later.

B. Content of the Work

In this monograph I will present the methods, results, and analysis of field work on bird communities of varying degrees of complexity in North and South America, with some reference to studies in western Europe. This work has been conducted more or less continuously since 1964, but the major effort was exerted in North American grasslands in summer 1964 and 1965; South American grasslands, fall 1965; North American scrub habitats, summers 1966, 1967, 1971, 1972, and in other seasons, 1967–1972; in South American scrub habitats, fall 1968 and 1971; and in Britain, summer 1970.

The major aim of these studies is to evaluate the extent to which coexisting species limit each others' activities and limit their common use of resources, and the extent to which they diverge (or converge) in their overall ecologies. The data are related to current ideas of niche and competition theory. Niche overlap, breadth, size, shape, and number of dimensions are discussed, as are some concepts of adaptive strategy such as "specialist" and "generalist," the notion of community stability, competitive release, optimal adaptation, ecological counterparts, and parallel and convergent evolution.

At 11 scrub habitat sites in North and South America the methodology followed and the sorts of results obtained are the same; these are the sites of "community studies" and provide the bulk of the data analyzed in Chapters 2 and 3. These sites, together with 10 simpler bird communities in grassland habitats, already described (Cody, 1968), encompass a fairly wide range of latitudes, altitudes, and habitat types: from 4″ high short-grass plains to 60′ pine-oak-juniper woodland. Using the results of these community studies, we analyze inter- and intracommunity variation in niche parameters. Ultimately a level of predictive sophistication is sought such that, given a large and varied species pool, an environment with known physical and structural properties, and a span of evolutionary time, the properties of the equilibrium community can be accurately estimated: numbers and relative abundances of the component species and their niche properties, i.e., the organization of the community.

This work is in no way intended to be a review; throughout most of the book, only the most pertinent references are included. Chapter 5, which deals with parallel and convergent evolution, was conducive to a broader treatment, as, to a lesser extent, was the subject of Chapter 6.

C. Organization of the Monograph

The remainder of this introductory chapter deals with the tactics of ecological isolation in birds and illustrates various "coexistence mechanisms" (Cody, 1968) which promote co-occupation of the same habitat by two or more species and their persistence there through time. These notions are then related to niche theory, and bring some perspective to the ill-used "competitive exclusion" principle.

Chapters 2 and 3 discuss the evolution of niche breadth

5

and the possible ways of generating a single niche overlap coefficient from its several components, and analyze the results of bird community studies with respect to variation in niche parameters. Niche breadth and overlap are related to the predictability of resources, to which the former shows little response and the latter varies dramatically in one component and only slightly (in the opposite direction) in the other components. The number of niche dimensions and the importance of niche shape are discussed, relative to the specialist-generalist concept, and the implications of niche overlap and breadth for species diversity are summarized.

In Chapter 4, I describe various situations of "competitive release"—alterations in the use of resources by species following changes in the competitive environment which constrains this use. The simplest case is character displacement (Brown and Wilson, 1956), and the most complex is the structure of new equilibrium communities on islands reduced in species numbers below their mainland counterparts. Chapter 5 reviews the evidence for the evolution of ecological counterparts between taxonomically unrelated species in structurally similar but geographically distinct habitats. These observations are extended to parallels in community makeup and organization so far as possible, with particular reference to the chaparral bird communities of Chile and California. Finally, in Chapter 6, exceptions to the normal patterns of ecological isolation are analyzed. Under conditions both of unusual abundance and unusual scarcity of food resources, bird species fail to evolve competitive displacement patterns. Further, in habitats of structure intermediate between those extremes preferred by two species, behavioral interactions evolve which regulate resource use between species exactly as normally occurs within a single species, by interspecific territoriality. By such behavioral devices, natural selection can reduce the number of ecologically distinct species to a value smaller than the number of taxonomic species in the community.

II. ECOLOGICAL ISOLATION AND COEXISTENCE MECHANISMS

In this section, I consider the various ways in which bird species differ from each other in their use of resources. The fact that species have evolved such differences has been recognized for a long time and was termed "ecological isolation" by Moreau (1948). Lack (1971) deals qualitatively with the subject and gives many examples. Ecological isolation is particularly orderly and precise, and therefore noticeable, in groups of related species, "guilds," which commonly show nonoverlapping ranges on a resource span such as a habitat gradient, food type, or vertical feeding range, but is equally easily identified, measured, and discussed in species of different genera or even families.

If ecologically similar species can be arranged in the form of a replacing series of partially overlapping distributions on some resources axis, they form a *displacement pattern*. The identification of such displacement patterns provides, first, circumstantial evidence for the ubiquity and importance of competition as a factor in community composition and organization and, second, points to the sorts of resources for which bird species compete. *Coexistence* is defined as the persistence of two species in the same habitat (or a 10-acre sample of the habitat), and such coexistence is achieved by the evolution of some minimal degree of difference in resource use. By feeding on partially different foods, by taking foods at different heights or feeding sites, by locating foods with different feeding behaviors, bird species can avoid competitive exclusion; these differences are called *coexistence mechanisms* and are discussed next.

A. Segregation by Habitat

1. Segregation on a Geographic Scale. Many pairs or larger groups of species occupy largely nonoverlapping or only partially overlapping geographic ranges and are called *geographic replacements*. Frequently such species show differences in their physiologies which correlate with differences in their respective ranges and in the environmental conditions which they encounter. Thus, in northern Europe, the yellow bunting *Emberiza citrinella* is a year-round resident and shows greater cold tolerance than the ortolan bunting *E. hortulana,* which, while it breeds at the same high latitudes, migrates south in winter (Wallgren, 1954). It would be a mistake to infer that such physiological tolerance differences are the cause of the geographic and/or seasonal limitation; rather, they are the result. Competition, through overall physiological, morphological, and ecological abilities, determines where the dominance of one species ends and that of the other begins, and thus reinforces any initially existing physiological differences.

Figure 1 shows the distributions of various bird species which serve as examples of geographic replacements. Precise replacement is shown in the three North American "species" of *Leucosticte,*[1] the rosy finches (Figure 1a). These near-species are probably in the process of further differentiation leading eventually to complete reproductive isolation. Likewise the longspurs *Calcarius* (Figure 1b), birds of short grassland and heath-tundra, show virtually no overlap between species and form a latitudinally replacing series. The fourth North American longspur, McCown's longspur, is placed in a different genus, *Rhynchophanes,* and is morphologically distinguished on the basis of its larger and thicker bill and its relatively shorter tail (Ridgeway, 1901). Although it extends further south than the chestnut-collared longspur *C. ornatus,* to Colorado, and is characteristic of shorter grassland, it shows extensive overlap in range with that species.

Figure 1c shows the distribution of two genera of tall- and mid-grassland finches, *Passerherbulus* and *Ammodramus,* each

---

[1] Peters, 1968, considered the three to be subspecies of a single holartic species, *L. arctoa.*

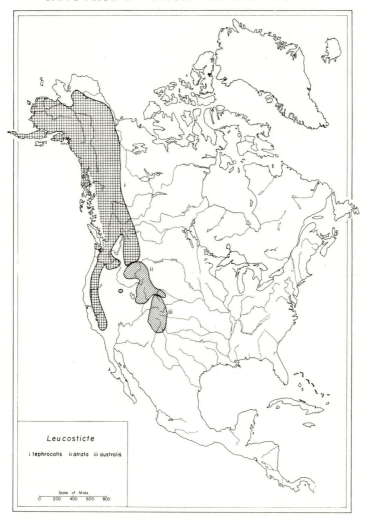

Leucosticte

i tephrocotis ii atrata iii australis

Scale of Miles
0   200   400   600   800

FIGURE 1a. Ranges of *Leucosticte* "semi-species," called by some authorities subspecies of the holarctic species *L. arctoa*.

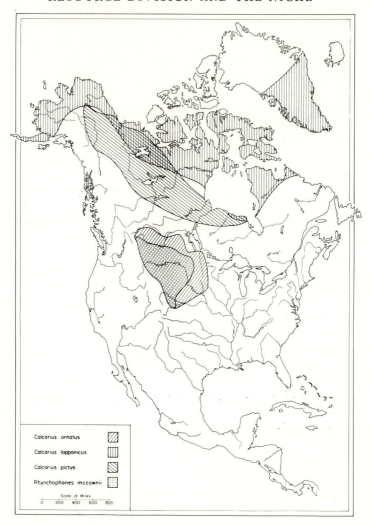

FIGURE 1b. Ranges of the four species of North American longspurs.

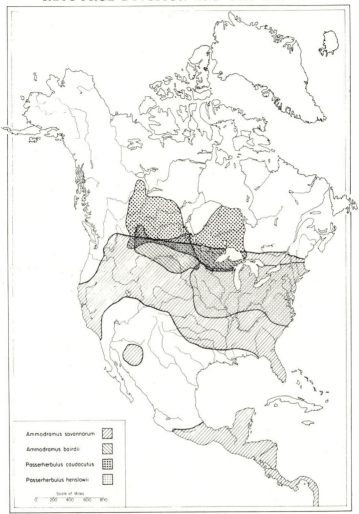

FIGURE 1c. Ranges of species in two genera of grassland sparrows.

11

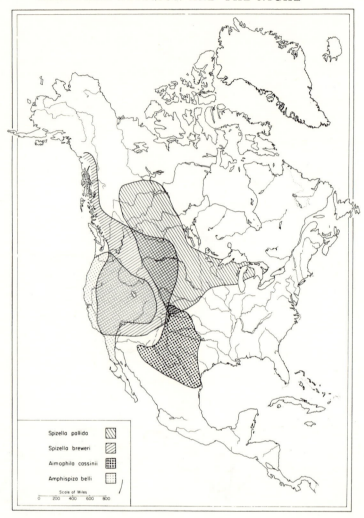

FIGURE 1d. Ranges of sparrows of sagebrush and saltbush habitats.

FIGURE 1e. Ranges of meadowlarks, with two congeneric largely allopatric species to the north and south, and a single species in a separate genus between them in tropical latitudes.

13

with two species. Henslow's sparrow, *P. henslowii,* is replaced in the northcentral United States and southcentral Canada by LeConte's sparrow, *P. lencontei,* with virtually no overlap. The closely related genus *Ammodramus* consists of the widespread grasshopper sparrow, *A. savannarum,* which overlaps Henslow's sparrow extensively and LeConte's sparrow marginally, and the more restricted Baird's sparrow. *A. bairdii,* whose range is almost entirely within that of its congener. Where these two species occur together in southern Saskatchewan, and perhaps elsewhere within their coincident ranges, they are interspecifically territorial (Cody, unpl.). Another widespread species, Savannah sparrow, *Passerculus sandwichensis,* occupies the northern half of North America, and overlaps *Passerherbulus* completely and avoids *Ammodramus* only in the southeast United States and in Central America.

The finches which are characteristic of low scrub such as sagebrush (*Artemesia*) or saltbush (*Atriplex*) are shown in Fig. 1d. Two congeneric species, clay-colored sparrow *Spizella pallida* and Brewer's sparrow *S. breweri,* replace each other across the United States and Canada over a northwest-southeast line, with some 300 miles of overlap in their range borders. These two species meet and show but marginal overlap with Cassin's sparrow, *Aimophila cassinii,* in the southern plains states, but the sage sparrow *Amphispiza belli* shows a considerable overlap with *S. breweri.* Finally the meadowlarks (Icteridae) are considered (Figure 1e). The two North American species, western meadowlark *Sturnella neglecta* and eastern meadowlark *Sturnella magna,* overlap in range on a southwest-northeast line from Texas to Ontario, a zone which ranges from 50 to 100 miles in width (Lanyon, 1957). In this zone of overlap, the species more finely subdivide habitat, with the western meadowlark predominating on the drier hilltops with shorter grass and the eastern meadowlark in the valleys with taller grassland. These differences are in accord with the average habitat each encounters in the center of its range (Cody, 1968). The two meet in intermediate habitat and are interspecifically territorial. Habitat more typical in structure of the western meadowlark, such as grazed fields, is occupied by the eastern meadowlark in the east, and the western meadowlark lives in Californian coastal marshes, more alike in structure to the habitat of the eastern bird. In South America the meadowlark replacements are two species in the Icterid genus *Pezites*,[2] which again are very largely nonoverlapping in geographic range, while the related but smaller *Leistes* species overlap extensively with both *Sturnella* and *Pezites* at low latitudes. The meadowlark role is played in Africa by pipits *Macronyx* (Motacillidae), and the convergent evolution these species show to their New World counterparts is discussed below (Chapter 5).

Within these species groups just discussed, the chief means of avoiding competition is a broad nonoverlap in geographic range. Two generalizations show up: (a) geographic replacement among subspecies, near-species, and species only recently differentiated is precise and without overlap. Ranges often show marginal overlap between the more distinct species in a genus, and overlap between species becomes more extensive as their genera show decreased affiliation; ranges overlap to the extent that the species involved are taxonomically unrelated. When closely related species within the genus do overlap extensively in geographic range, such as *Sturnella* and *Ammodramus*, and almost certainly *Spizella*, competition is mediated between species by dint of interspecific territoriality. (b) Just as obviously, geographic replacements show a more distant taxonomic relationship as the geographic areas which support the species' habitat are more widely separate. Thus the meadowlarks are replaced locally by different subspecies of *Sturnella magna*, e.g. *S.m. argutula* by *S.m. magna* between the southeastern and the northeastern United States, *S. magna* is replaced by *S. neglecta* between the eastern and western

[2] Short (1968) has suggested merging *Pezites* with *Sturnella*.

15

United States, *Sturnella* by different genera in the same family Icteridae, *Leistes* and *Pezites*, between North and South America, and by different families, Icteridae by Motacillidae, between the western hemisphere and Africa.

2. Segregation by Altitude. Bird species may be separated in their altitudinal ranges; this is especially common, of course, on steeper mountainsides. As this type of segregation is somewhat intermediate between geographic and habitat segregation, it is worth treating separately. Related species, often congeners, may replace each other in sequence along a transect up an elevational gradient. Diamond (1972) has published results of altitudinal surveys of birds in New Guinea mountains, and finds that as many as four species may occur in these sequences. Terborgh (1971) has shown that similar series occur on the eastern slopes of the Peruvian Andes, and here again up to four species divide the vertical range between them. Although around 50% of the 261 species occupied altitudinal ranges not obviously influenced by competitors but possibly by physiological tolerances, and a further 19% were bounded by vegetational changes, the remaining one-third of the species censused formed altitudinally replacing series and qualify as strict altitudinal replacements.

An interesting point about these data, for both studies were made in tropical locations, is that virtually no overlap occurs between adjacent members of the replacing series; in fact the species turnover may be as precise as the line which marks the edge of a species territory- in spite of its being apparently unrelated to habitat turnover. Two lines of evidence bear on the habitat independence of the species turnovers. First, the elevations at which species turnovers occur are generally different in different species sets on the same mountainside; second, and more important, the elevations at which turnovers occur may be at different heights in the same species set on different mountains.

But there must be a feedback from the character of the vegetation into the mechanisms of competition, even though the vegetation varies clinally with few if any abrupt changes along the altitudinal transects. These displacement patterns are certainly products of interspecific competition, as Diamond has shown that, in the absence of some members of a series, the adjacent members adjust their elevational ranges to occupy the vacant range.

These abrupt species turnovers are in striking contrast to some altitudinal replacements observed on temperate mountainsides. Several examples can be cited from the Sierra Nevada and San Gabriel Mountains in California, in which apparent altitudinal replacements show considerable overlap in range. Starting with the lower species, this occurs in the titmice *Parus inornatus* and *P. gambeli*, the jays *Aphelocoma coerulescens* and *Cyanocitta stelleri*, the sapsuckers *Sphyrapicus varius* and *S. thyroideus*, the finches *Carpodacus mexicanus* and *C. purpureus*, the thrushes *Hylocichla ustulata* and *H. guttata*, and others. In these species pairs, the extent of altitudinal overlap in range may be as much as 50% of the total range of either. Although little is known of any agonistic interactions between species where they overlap, these may be of common occurrence, and perhaps even necessary for the combined occupancy of the habitat. Some evidence exists for the jays, which are interspecifically aggressive (B. Bundick, pers. comm.) where they meet.

Altitudinal distributions were plotted in the southern temperate Andes by Cody (1970). The genus *Muscisaxicola* is represented by six species at the latitude of Santiago; while each species is characteristic of a certain elevational range less than the total available to the genus (10,000'), there are such extensive interspecific overlaps that at least as many as four species may be found together at intermediate altitudes. Here, there were clearly interspecific behavioral interactions among species, such that space and the resources it contained was

1 7

divided among species. Normally this division occurs only within species. The species of at least one other genus, *Geositta,* shows similar altitudinal range overlaps. Altitudinal segregation is apparently less viable a coexistence mechanism in temperate than in tropical localities, for reasons possibly associated with resource predictability and discussed later.

3. Local or Between-Habitat Segregation. It is a common observation that related species differ in their preferred habitats. At one extreme the segregation is so complete that not more than one species of a taxonomic group may be found in any one place. The thrashers, Mimidae, are an example of this (Figure 2). Each species occupies some variety of scrub

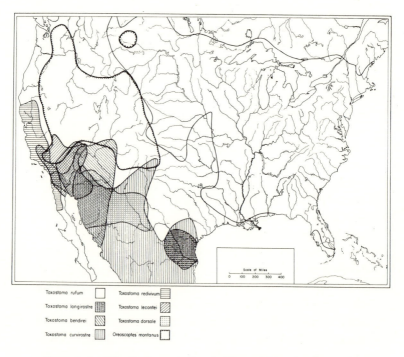

Toxostoma rufum
Toxostoma longirostre
Toxostoma bendirei
Toxostoma curvirostre
Toxostoma redivivum
Toxostoma lecontei
Toxostoma dorsale
Oreoscoptes montanus

FIGURE 2. Distribution of the habitat-specific thrashers of North America.

18

habitat, and the usual situation is that a single species is found in a homogeneous vegetation type (e.g., chaparral, Mohave Desert). Inasmuch as their species-specific habitats are geographically distinct, the seven North American thrashers might be called geographic replacements. Only in southwestern parts of the country, such as near Portal, Arizona, where the habitats of four species abut, can one find several species in a half-mile walk.

There are five species with long bills typical of taller scrub and two short-billed species that live in drier and shorter vegetation types. Six of the seven species are in the genus *Toxostoma;* there is a tendency for the only short- and straight-billed of these, *T. bendirei,* to show some overlap with the long- and curve-billed species, in particular with *T. dorsale* and to some extent *T. lecontei.* I have found the second short-billed species, sage thrasher *Oreoscoptes montanus,* living in the same habitat (eastern Mohave Desert) as *lecontei.* Other mimids, such as the mockingbird *Mimus polyglottos* and the catbird *Dumatella carolinensis,* which are more arboreal and possess much shorter, straight bills, commonly occur in the same habitat as *Toxostoma* representatives.

In the emberizine finches (recently removed from Fam. Fringillidae into Fam. Emberizidae), segregation by habitat is also a rule, although a particular 10-acre patch can hold several species. There is a vegetation type with particular properties of foliage height and density characteristic of each ـpecies and more or less distinct from the vegetation types preferred by others. Figure 3 plots the height of vegetation against the "half-height," the height at which half the vegetation density or leaf area is above and half below. This information comes from a plot of vegetation density against vegetation height in the "foliage profile" (Cody, 1968). On this graph the habitat ranges of the North American emberizine finches are distributed and generally well-spaced. In some, the depicted habitat ranges are accurate, as I have measured the

19

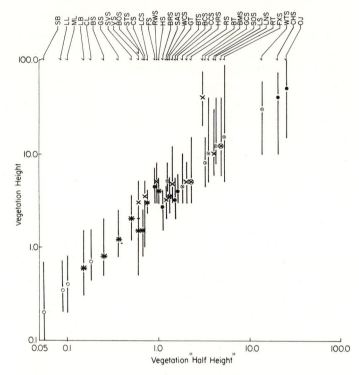

FIGURE 3. Distribution over habitats of the North American emberizid finches. Species are initialed across the top, and their average (dot, cross) and range (vertical bar) of habitat use is shown relative to vegetation height and "half-height," the height at which half the vegetation density is below and half above. Open circles: longspurs and snow bunting; stars: six genera of strictly grassland sparrows; crosses: desert sparrows *Aimophila, Amphispiza,* and *Chondestes;* dots: woodland and forest *Spizella* and *Junco;* crossed circles: towhees; dotted circles: scrub and undergrowth sparrows of *Passerella, Melospiza* and *Zonotrichia.* SB = snow bunting, LL = lapland longspur, ML = McCown's longspur, LB = lark bunting, CL = chestnut-collared longspur, BS = Baird's sparrow, GS = grasshopper sparrow, SVS = savannah sparrow, VS = vesper sparrow, BOS = Botteri's sparrow, STS = sharp-tailed sparrow, CS = Cassin's sparrow, LCS = LeConte's sparrow, FS = field sparrow, RWS = rufous-winged sparrow, HS = Henslow's sparrow, BRS = Brewer's sparrow, SAS = sage sparrow, WCS = white-crowned sparrow, GT = green-tailed towhee, BTS =

2 0

preferred habitat type in many different parts of the country. In other species, I have fewer firsthand data, only enough to approximately position the species. A logarithmic scaling of axes in Figure 3 is necessary to produce approximate equal habitat use areas for the finches. This brings up problems of niche breadth and scaling that are discussed below.

The number of coexisting species in a study plot is thus largely a function of the vegetational diversity included within its bounds. Cody (1966) found that natural grassland areas of around 10 acres in size and selected for their homogeneity usually supported one or two emberizine finches (plus one or two species of other families to bring the passerine total to three or four), but this number can certainly be increased. Wiens (1969), for example, studied an 80-acre Minnesota grassland with four finch species (out of seven passerine species) in residence, and the number can be further enhanced by inclusion within the census area of topographic, ecotonal, and man-made influences. In 1966 I censused grasslands in Buffalo Pound Provincial Park, southern Saskatchewan, by walking a half-mile transect across a) natural mid-grass prairie, b) sown Russian rye, c) mowed natural prairie (Figure 4). The species list included five emberizine finches in the grasslands, and an impressive total of nine grassland bird species, considerably more than is to be expected in undisturbed natural grasslands. A species-area curve for the breeding passerine bird species of North American grasslands, centered in Minnesota, is given in Figure 5. As area is increased on a local scale, bird species are accumulated rapidly, a reflection of the addition of new species as new habitats are

---

black-throated sparrow, BCS = black-chinned sparrow, CCS = clay-colored sparrow, HRS = Harris' sparrow, RS = rufous-crowned sparrow, BT = brown towhee, BMS = Bachman's sparrow, GCS = golden-crowned sparrow, SOS = song sparrow, LS = lark sparrow, LNS = Lincoln's sparrow, RT = rufous-sided towhee, FXS = fox sparrow, WTS = white-throated sparrow, CHS = chipping sparrow, OJ = Oregon junco.

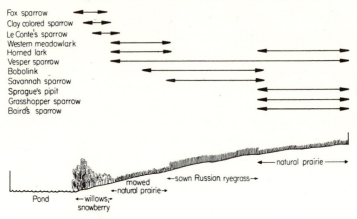

FIGURE 4. Habitat preferences of grassland birds censused in Buffalo Pound Provincial Park, southern Saskatchewan.

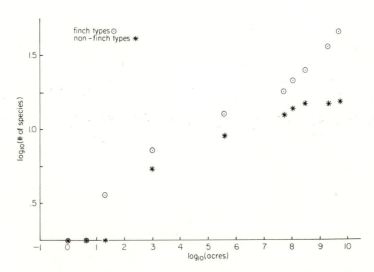

FIGURE 5. Species-area curves for grassland birds, beginning in tall-grass prairie in west-central Minnesota.

included into the survey area. Numbers of species continue to increase with survey area even after the great part of the range of habitats is included. This continued increase in species numbers reflects the inclusion of geographic replacements in the species list. Addition of breeding bird species with increase in local area reflects a turnover of species with habitat; Figure 5 shows that this turnover rate is similar in the finches and non-finches alike. However, even after a wide range of habitats is included (Minnesota supports tall-grass prairie in the center of the state near the deciduous forest ecotone and short-grass plains habitat in the Red River Valley), finch species continue to accumulate, at least as rapidly as before, while non-finches, chiefly icterids, remain almost constant. The larger and wider-ranging non-finch species are habitat-specific, but not locality-specific: the finch species change with both habitat and geographic area.

Both the partially grazed area studied by Wiens and the Buffalo Pound site fall above this species-area curve. So do census data from natural grasslands in the Kalahari region of (tropical) South Africa (Gordon Maclean and M. P. Stuart-Irwin, pers. comm.), in apparent contradiction to my earlier suggestion (Cody 1966) that grasslands may support similar levels of bird species everywhere.

4. Within-Habitat Segregation. For the first time, I now consider ways competition is avoided in species which by definition are actually coexisting, that is, living together in the same patch of uniform habitat and co-occurring in the 10-acre census areas I use as study sites. Bird species which co-occur in the same patch of habitat may separate their ecological activities in several ways, and in this section a horizontal separation of feeding sites is discussed. Feeding activities may be confined to a) different parts of trees or bushes, b) bushes different in those structural aspects that influence the foraging efficiency of a particular phenotype (e.g. different species of plant), or c) different sections of the habitat characterized by overall differences in vegetation structure.

The first of these possibilities was shown by MacArthur (1958) to distinguish the foraging sites of *Dendroica* warblers that live together in Maine spruce forests. This sort of microhabitat displacement may be important in the coexistence of large numbers of congeneric species, but is not commonly encountered in bird species. In the communities I studied, the only examples found were in the nuthatch genus *Sitta,* in which the smaller species *S. pygmaea* feeds further out along pine branches than does the larger *S. carolinensis* in Arizona pine-oak-juniper forest, and possibly also in the towhee genus *Pipilo,* where *P. erythrophthalmus* feeds on the ground in deeper litter under bushes and *P. fuscus* feeds in the same manner further out from bushes in California chaparral.

Foraging individuals of a particular species have the option of visiting every piece of vegetation in their path, but may preferentially select some types, such as trees of a particular species, and avoid others. This might be expected to occur only in habitats of discontinuous vegetation, such as woodlands and deserts, composed of more than one common size or species of plant. This again does not seem to be a common pattern, and I have come across only one potential example. The pine-oak-juniper woodland occupies a position on the sides of Arizona mountains below the continuous ponderosa pine forest and above the oak woodlands and oak grasslands. The black-throated gray warbler, *Dendroica nigrescens,* occupies the pine-oak zone and also the oak woodland below; Grace's warbler, *Dendroica graciae,* likewise occurs in the pine-oak and lives also in the pine forest above. Where these two species occur together in the pine-oak, the former appears to be an oak (and juniper) specialist, and the latter a pine specialist. Both species occassionally use trees of the nonpreferred type, but tend to avoid them (Marshall, 1957, and pers. obs.).

The third possibility for within-habitat segregation is an important way in which coexisting species differ from each other, and has been the subject of considerable measurement and study. There is often a tendency, even within quite uniform habitats, for different species to select quite different parts of a habitat patch in which to feed and raise young. Figure 6

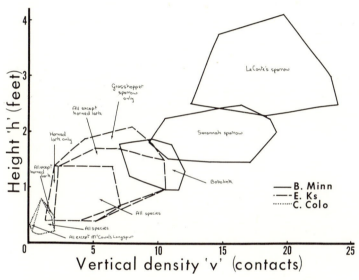

FIGURE 6. Differential habitat preference of birds in three North American natural grasslands. From Cody, 1968a.

shows bird habitat distributions within three North American grassland communities: short-grass plains in Colorado with four passerine species, mid-grass prairie in Kansas with three species, and tall-grass prairie in Minnesota, also with three species. Although each of the three study areas is about of equal extent, there is practically no within-habitat displacement in the Colorado grassland; on the other hand, there is scarcely any habitat overlap among the Minnesota species, and the Kansas species are intermediate in this respect.

The within-habitat segregation observed in Minnesota grassland occurs to a greater or lesser degree in every habitat. In some climatically predictable and homogeneous vegetation types such as chaparral and desert, the average within-habitat overlap between pairs of resident bird species, designated α_H, often exceeds 70%, but in other habitats is often far less and is correspondingly more important in the separation of ecological activities. Figure 7 illustrates the habitat heterogeneity pre-

FIGURE 7. Habitat heterogeneity in superficially uniform willow habitat in Grand Teton National Park, showing different vegetation heights. Birds select and use rather different habitats within this plot.

sent in a superficially rather homogeneous habitat, willow scrub (*Salix*) in Jackson Hole, Wyoming. The study area consists of ninety-eight 50′ × 50′ squares, each contoured with

26

respect to vegetation height from grass ($\pm 1'$) to 14' in intervals of 2'.

This willow habitat is occupied by five finches, three to four warblers, a thrush, flycatcher, magpie, and a hummingbird.[3] Each selects parts of the habitat which are distinguished in some way or another from the rest of the area. To show that this is so, I obtain the mean vegetation profile of the habitat occupied by each species and compare it to the mean profile of the willow habitat as a whole. Thus the horizontal vegetation density at a particular height at a randomly selected point in the territory of one of these species deviates by a certain amount from that at a point selected at random from anywhere within this habitat. Figure 8 gives the habitat preference of the warbler and finch species of the willows in terms of $\pm \%$ deviation of vegetation density in territories from that in the habitat in general, as a function of height above the ground.

There are clearly recognizable trends which distinguish the species' habitat preferences. Among the warblers, the yellow-throat prefers a habitat denser at the 0–2' level than the average patch, the yellow warbler prefers that which is denser at the 3' and 6–9' levels, and Wilson's warbler occupies the tall habitat with canopy at 9–12'. MacGillivray's warbler, *Oporosnis tolmiei*, also occurs in willows in Jackson Hole but prefers even an taller habitat around 12–15' high and was of marginal occurrence in this study area. The five finch species show an orderly gradient of habitat preference from clay-colored sparrow around the grassy patches and away from the taller shrubbery to white-crowned sparrow, which avoids the low vegetation and selects the taller habitat. The finches occupy generally lower vegetation than the warblers. Figure 9 gives smoothed curves that summarize the relations.

Within-habitat overlap can be measured between species

[3] Further data on community study sites are given in Appendices A and B.

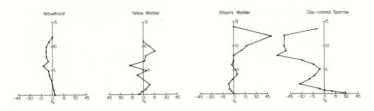

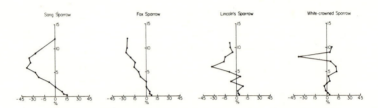

FIGURE 8. Habitat selection in eight species of birds in Wyoming willows. The habitat within defended territories is expressed as a frequency distribution of vegetation height in each bird species, and this frequency distribution is compared to and deviates from the frequency distribution of vegetation heights in the study area as a whole in the manner shown.

pairs by simply dividing the area of habitat occupied by both species by the geometric mean of the areas occupied by each species separately. This gives a value of α_H between 0 and 1. Alternatively, the vegetation characteristics can be mapped and the habitat overlap calculated as before, for example from areas on a plane of two habitat variables as in Figure 6. The former is more conveniently measured, and α_H values as simple spatial overlaps are used as indices of habitat overlap. Exceptions to the assumption of a direct equivalence of spatial and habitat/vegetation overlap are given in Chapter 6.

B. Vertical Stratification of Feeding Zones

Species that co-occur at a point in the habitat can still separate their activities in space by a vertical segregation of feeding

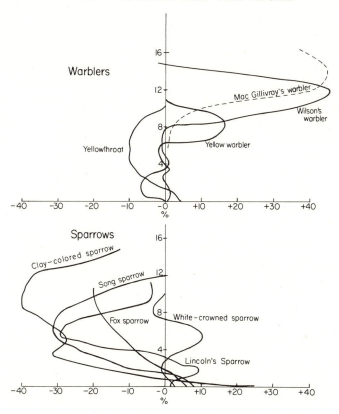

FIGURE 9. Habitat selection in warblers and sparrows of Wyoming willows.

zones. Species which forage at different heights above the ground will come into competition to the extent that these feeding zones overlap. Not surprisingly, this coexistence mechanism is unimportant in short- and mid-grasslands below about 3' in height (Cody, 1968), but in all other taller habitats it is the single most important factor in the segregation of species' feeding activities.

A vertical stacking of feeding zones among species has often

29

been recognized in bird community studies. For instance, Gibb (1954) studied coexistence in the titmice (Paridae) of England and showed that different species feed at different heights in habitats where several occur together. MacArthur and his associates (1965, 1966, and earlier) found that, in a single geographic area, bird species diversity is well-correlated with the vertical structure of the habitat. Further, a large part of the latitudinal gradient in bird species diversity is accounted by the realization that birds apparently respond to three vertically stacked horizontal layers in temperate habitats (a grass-forb layer 0–2', an understory layer 2–20', and a canopy layer >20'), but to four such layers in tropical habitats (0–2', 2'–10', 10–25' and >25'). This conclusion was suggested by bird census and habitat diversity data, and supported by limited observation on the feeding heights of the species censused. Karr (1972) has accumulated extensive data on vertical foraging ranges in Illinois and Panama, and can document the greater restriction of the tropical birds. On a species-poor island, Puerto Rico, vertical foraging ranges were expanded, and species diversity was predictable on the basis of a division of the total vertical foraging range into two layers (Figure 10, from MacArthur *et al.,* 1966). More recently MacArthur *et al.* (1972) have shown similar vertical foraging range expansions on species-poor islands in Panama Bay, in comparison to mainland Panama ranges.

The vertical foraging distributions of birds are measured by timing feeding individuals with a stopwatch and estimating foraging heights. In order to obtain the most complete picture, short bouts of continuous feeding activity are timed for many different individuals in different parts of the habitat and at all times of the day. Data are collected on a particular individual in a particular location up to a limit of 60 seconds, in order to compile species feeding height distributions from a broad sample of individuals, times, and locales. At least 1000 seconds of continuous uninterrupted feeding height data were

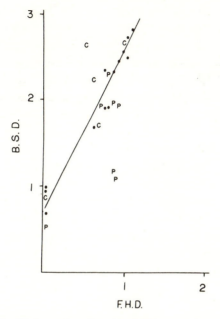

FIGURE 10. Bird species diversity B.S.D. is plotted against foliage height or habitat structure diversity F.H.D. in censuses from North America (dots), Panama Canal Zone (marked C) and Puerto Rico (marked P). From MacArthur *et al.*, 1966)

collected per species, which entailed an average 30 feeding sequences per species.

It was evident from the first that bird species recognize and restrict their activities to much more finely divided height intervals close to the ground than in the canopies of trees. Some species spend all of their time foraging on the ground; others feed between the ground and 3′ above it, but no species is similarly restricted to a 3′ interval in the canopy. Generally, the range of feeding heights increases with the mean foraging height above the ground. When a plot is made of height interval versus proportion of time spent foraging within that height interval, the height axis must be scaled to take account of

31

this nonlinear subdivision. In tall habitats the height intervals used are <ground, ground, ground–6″, 6″–2′, 2–4′, 4–10′, 10–20′, 20–35′, 35–60′, and >60′; extensive observation suggests that a species could reasonably be expected to confine most of its feeding activity to one of these zones, but not to a smaller interval at that elevation above ground. In lower vegetation, this scale is too coarse, and finer-subdivisions are used.

An example of bird feeding height distributions is given in Figure 11, for the birds of the Teton willows whose habitats are discussed above. Song, fox, white-crowned, and Lincoln's sparrows are ranked in this order from lowest and mostly ground-feeding to uppermost in the first three feet of the foliage, and the three warblers and clay-colored sparrow are

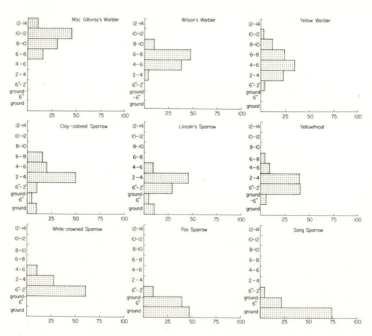

FIGURE 11. Distribution of foraging heights for nine bird species of the Wyoming willows.

ranked from the lower yellowthroat up to the higher Wilson's warbler at around 6–10′ above the ground. Where MacGillivray's warbler is present, it forages on average higher than Wilson's warbler. The overlap between any pair of these species in vertical foraging distribution is simply the area common to their two distribution curves, denoted α_V and ranging from 0–1.0.[4]

Segregation by vertical feeding heights is particularly common in foliage insectivores and sallying flycatchers. It also frequently distinguishes aerial insectivores, such as swallows and swifts, as species of the former often feed within a hundred or so feet of the ground but the swifts customarily higher. Further examples of such vertical feeding segregation are shown for pairs of bird species from an Arizona pine-oak woodland (site characteristics in Appendix A), species which but for this tendency to forage at different heights would be ecologically extremely similar. In a sense these species are ecological counterparts at different heights in the vegetation. Figure 12a illustrates feeding height distributions in a pair of aerial insectivores, violet-green swallow and white-throated swift; the canopy insectivores, Grace's and black-throated gray warblers; and the sallying flycatchers, western wood peewee and olivaceous flycatcher. Other important ecological differences, such as size and feeding behavior, further distinguish the species pairs, but segregation by feeding height is of primary importance. The mean vertical overlap α_V between the three pairs is $(0.50 + 0.27 + 0.18)/3 = 0.32$. A further example of this pattern is given later, in Figure 61, for those who wish to look ahead.

In contrast, vertical overlap is much greater in other groups of ecologically similar species such as the trunk insectivores,

---

[4] Schoener (1968) has written this overlap index as

$$\alpha_{xy}(D) = 1 - \tfrac{1}{2} \sum_{i=1}^{n} |P_{x,i} - P_{y,i}|$$

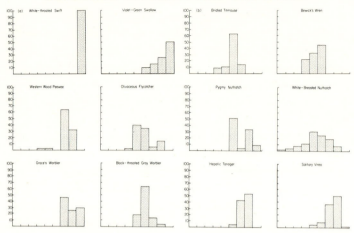

FIGURE 12. Distributions of foraging heights for six pairs of ecologically related bird species of the Arizona pine-oak. Height intervals along the abscissa are ground, ground–6″, 6″–2′, 2′–4′, 4′–10′, 10′–20′, 20′–35′, 35′–60′, and above 60′.

branch-and-twig insectivores, and the slow-moving and steadily-searching canopy insectivores such as vireos and tanagers (Figure 12b). These species are distinguished from the others discussed above by their higher "search/pursuit" ratios, that is, they spend relatively more time searching for prey and less time pursuing it. In the same pine-oak woodland, examples come from two species of trunk foragers, the pygmy and white-breasted nuthatches; the branch-and-twig insectivores, Bewick's wren and bridled titmouse; and the slow-searching foliage insectivores, solitary vireo and hepatic tanager. Among these three species pairs, vertical overlaps α_V average $(0.64 + 0.35 + 0.90)/3 = 0.630$, and correspondingly greater emphasis is laid on other coexistence mechanisms.

C. Differences in Food and Feeding Behavior

Species which find themselves at the same point in space, that is $\alpha_H = \alpha_V = 1.0$, can still differ in their resource use by using different food items. This can be so because a) they may encounter different food items, and/or b) they may be morphologically equipped to eat different sizes, shapes, or hardnesses of food items.

It was earlier recognized (Cody, 1968) that mere analysis of stomach contents can give an extremely biased picture of the ecological overlap between species. Stomachs can show a great deal of diet overlap, and yet the two species might be feeding in different ways or in different places such that each is harvesting food not available to the other. Less likely is the possibility the ecological overlaps might be underestimated; yet species that show different stomach contents might be generally more similar than is apparent, if their gut contents are biased by the times of day they are sampled and foraging sites recently visited. I have therefore adopted the approach that a knowledge of feeding behavior and of bill morphology is necessary and sufficient to measure the food and feeding aspects of ecological isolation. Feeding behavior determines which food items will be encountered, and bill morphology determines which of those encountered will be accepted and incorporated into the diet. As bird species seem to be largely opportunistic in what they feed on, every food item encountered and found to be manageable will be incorporated into the diet, and none such is likely to be passed over.

The evidence that bill structure is important in determining the diet (and, of course, vice versa) comes from a number of sources. It must be summarized here because of the importance I place on it as an index of ecological similarity. Finches in particular have been studied. Lack (1947) associated bill differences to differences in diet of the Geospizinae of the Galapagos Islands, and concluded that the adaptive radiation of bill sizes and shapes in these finches was effected through competition for food. Bowman (1963) has quantified the relation between size and shape of bill in four *Geospiza* species and the size and hardness of the seeds they eat, and has shown that bill size in the insectivorous genera *Certhidea, Camarhynchus,* and *Cactospiza* is related to the size of the insect larvae they feed on. In all cases the relationships are those expected; large food items with larger bills and hard food items with deeper bills. Kear (1962) investigated British finches, which differ chiefly in bill morphology and show a great deal of overlap in habitat. Differences in preference of seeds of various sizes were correlated, in the predicted way, with bill structure and the efficiency with which the birds could husk the seeds. These same finches were studied by Newton (1967), who found that, while the small-billed species were restricted to smaller seeds, the large-billed species took a wide range of seed sizes, including small and large, and showed preferences for the larger seeds. Two North American finches also were tested in seed-choice experiments (Hespenheide, 1966) and found to differ in their seed-size preference and husking efficiency in the expected way. Further experiments by Willson (1971) demonstrated the same divergence in preference, and unpublished data of R. H. MacArthur and D. MacArthur for Sonoran Desert finches confirm the picture.

Coexisting species of terns (Laridae), herons (Ardeidae), and accipiter hawks (Accipitridae) differ chiefly by body size, and in each case the size differences have been shown to be related to differences in prey size distribution (Ashmole and Ashmole, 1967; H. Recher, unpublished; Storer, 1966). An important recent paper by Hespenheide (1971) discusses the phenomenon, in particular the sizes and diet relations of the flycatchers Tyrannidae. This paper makes two valuable points. First, the prey size distributions of insectivorous birds are log-normal, a result which parallels the observation that the recognition of and restriction to vertical foraging layers approach normality only when foraging height is scaled in some power function (see above). Second and central to this discussion, coexisting species of different bill and body sizes differ in the mean size of insect prey taken. Characteristically,

36

forest and scrub habitats in eastern North America support two tyrant flycatchers, one large and one small. This is true also for the Arizona pine-oak woodland, the Californian chaparral, and for Chilean matorral (see Chapter 5). The prey size distributions of two pairs of flycatchers which are commonly found together in the eastern United States are shown in Figure 13; this segregation by prey size is further

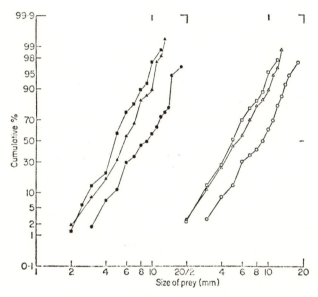

FIGURE 13. Size frequency distributions of beetle prey taken by flycatchers; species are eastern kingbird (circles), eastern phoebe (triangles) and Acadian flycatcher (squares). Open symbols are cumulative frequency by species. From Hespenheide, 1971.

exemplified in swallows and swifts (Hirundinidae and Apodidae) and vireos (Vireonidae), and is presumably of widespread occurrence (cf. Schoener and Gorman, 1968, for data on the head sizes of *Anolis* lizards and their insect prey distributions).

37

The possibility exists that body size is a more accurate predictor of prey size distribution than is bill size in birds. Schoener (1965) suggests that body size ratios might be more important indices of character difference in raptorial birds, and Lack (1971) ventures the same opinion with reference to island species of white-eyes (Zosteropidae) in the Indian Ocean. It is not clear, a priori, when body size ratios are not directly proportional to bill size ratios, which index is the more ecologically meaningful as a measure of probable diet dissimilarity; the evidence is equivocal. Bill size may be less subject to selection pressures other than those of competing species with diet overlaps, and is likely therefore to reflect more consistently the results of competition for food. Bird species involved in character displacement appear to show more sensitivity to bill size than to body size displacement, and the same is true for the Chilean species of character convergence situations (See Chapter 6). The woodpeckers *Centurus* on Puerto Rico show increased sexual dimorphism in the absence of competing species—one aspect of competitive release—and differ more between the sexes in both body and bill sizes; however, the percentage of difference between the sexes increases from 10.5 to 18.1 between mainland and island in body weight, but from 9.1 to 21.3 in bill length (Selander, 1966). Hespenheide (1971) showed that both body size and bill size are well-correlated to the mean prey size of the insectivores he studied, but that in general the correlation was higher for body sizes than for bill sizes; the correlation coefficients were not, however, significantly different from each other. In arctic sandpipers of the genus *Calidris,* bill size is a much more accurate indicator of prey size difference than is body size (Figure 14) in the four species studied by Holmes and Pitelka (1968). It seems from the available information that ground-feeding birds such as *Muscisaxicola* and the sandpipers show a more orderly serial array in bill size than canopy feeders, such as the pigeons and lorikeets studied by Diamond (1972), where

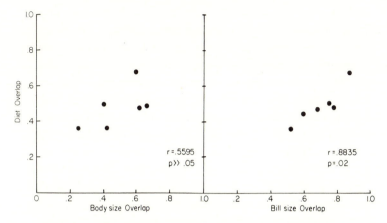

FIGURE 14. The relation between diet overlap and a) body size similarity, b) bill morphology similarity in arctic waders. From data in Holmes and Pitelka, 1968.

size may be more relevant. Smaller twigs support only the smaller, lighter species, and exclude the heavier birds. Clearly evidence exists to support either view and relates bill size to prey size sufficiently strongly that we may be comfortable with either measure.

As indices of similarity in bill structure between species pairs, I use a) difference in bill length l: $\alpha_{bl} = 1 - |l_1 - l_2|$ for $|l_1 - l_2| < 1''$, and $\alpha_{bl} = 0$ for $|l_1 - l_2| \geq 1''$, and b) difference in the ratio of bill depth d to bill length l: $0 \leq \alpha_{bd} \leq 1.0$ where $\alpha_{bd} = 1 - |d_1/l_1 - d_2/l_2|$.

Feeding behavior, while recognized by some ecologists as an important variable in which bird species can differ and therefore reduce ecological overlap in diet, has seldom been quantified. MacArthur (1958) incorporated this factor into his study of warbler coexistence, which was perhaps the earliest effort to measure this variable. I have shown (Cody, 1968) how feeding behavior can be easily quantified by a plot of a feeding sequence, distance against time, that shows the mean feeding "steps" of an individual. The usual feeding sequence is a progression involving repeated stops and starts. Characteristic of each species' feeding behavior is a) an average speed of progression (total path distance traveled/total time taken to traverse that path), b) the mean duration of a "stop" in the sequence (the total time stationary/number of stops made), and c) the mean percentage of time the bird is stationary (total time stationary $\times$ 100/total time of the feeding sequence). These three parameters characterize a "saw-tooth" curve which is a picture of species-specific feeding behavior. The data from which the averages of the feeding behavior parameters come are collected at the same time and with the same restrictions as the information on feeding height distribution. The raw data entered into a notebook are a) total time T_T seconds the bird is watched, measured on a stopwatch, b) total time T_S seconds the individual was stationary during the feeding sequence, accumulated on a second stopwatch; c) the total number n of stops made during the feeding sequence, counted in the observer's head. When the feeding sequence ends for one reason or another, or the observer terminated his watch on it, d) the path distance x the bird has traveled is retraced mentally and estimated in feet. Thus average speed $v = x/T_T$ ft sec, duration of average stop $s = T_S/n$ and % time stationary $t_s = T_S \times 100/T_T$.

Examples of feeding behavior curves are shown in Figures 15–17, and some obvious trends can be pointed out in these data. The sparrows in the Wyoming grass-sage area (see Appendix A) show that species which feed on the ground move more slowly and spend more time stationary, searching for food in one spot (Figure 15). The two species which feed mostly off the ground in the sagebrush—white-crowned sparrow and Brewer's sparrow—move more rapidly and make fewer and shorter pauses. Both vesper sparrow and savannah sparrow feed on the ground, the former also between the ground and 6″ above it; the vesper sparrow avoids the dense

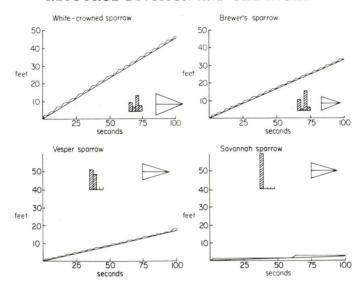

FIGURE 15. Quantitative representation of bird feeding behavior, from averaged steps in feeding sequences. Bill sizes and shapes to same scale. Histograms represent feeding heights, proportions of observations within the intervals (from left) ground; ground–6″, 6″–2′, 2′–4′.

grass areas preferred by the savannah sparrows, and moves more rapidly through the sparse, open grassy areas among sagebrush. Both velocity and duration of average stop are related to the density of the vegetation in which birds forage.

The sparrows of the Teton willows all maintain more or less the same average speed of progression (Figure 16), but are distinguished from each other in the duration of average stop (the stadia lengths in the feeding behavior curves). Again this appears to be most likely related to vegetation density. The warblers spend less time stationary and move more rapidly through the foliage. Wilson's warbler darts from branch to branch, plucking insects from the foliage and out of the air as well as searching for them on the leaves. This active feeding behavior is permitted by the low density of the vegeta-

41

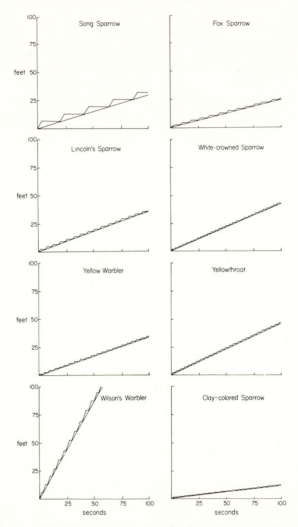

FIGURE 16. Feeding behaviors of birds of the Wyoming willows.

tion it searches through; vegetation densities are higher both below and particularly above its preferred feeding height (Figure 8).

Finally, the Arizona pine-oak insectivores are shown in Figure 17, and, between them, they show a variety of feeding

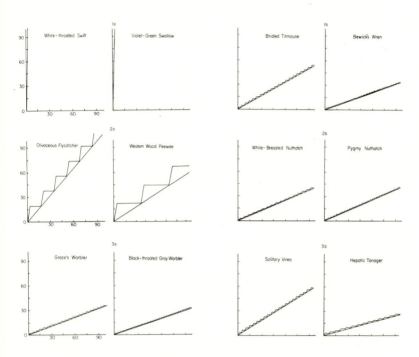

FIGURE 17. Feeding behaviors of birds in the Arizona pine-oak.

behavior patterns. The aerial insectivores do not, of course, spend time stationary while feeding; swifts average around 50 ft sec and do not descend to canopy height, whereas the swallows average 30 ft sec and spend some time feeding between the taller trees. The feeding curves of the two flycatchers reflect their habit of sitting and waiting for insect prey to fly by; most of their time is spent stationary, and

43

movements from the perch after prey are rapid darts. The two warblers spend similar proportions of their time in motion, and progress with similar average speeds; Grace's warbler feeds mostly in pines with a much more open foliage than the oaks and junipers preferred by the black-throated gray warbler, makes correspondly longer flights between stops, and searches longer at any one site.

There are four twig, branch, and trunk searchers in the pine-oak, and these show a considerable range of average speeds. Even more distinctive are their differing tendencies to search while on the move—low value of t_s, such as 32.4% in white-breasted nuthatch—versus stopping to search minutely in some crack or nook, such as bridled titmouse, with $t_s = 70.9\%$. The two nuthatches have a similar, intermediate average speed of progression, $v = 0.40$ and 0.41 ft sec, and their mean stop lengths are similar at $s = 1.87$ and $s = 2.10$ sec, but they are distinct in t_s values of 32.4% versus 61.3%. Lastly, the two slow-moving insectivores, hepatic tanager and solitary vireo, have the same style of feeding in terms of t_s (81.3% vs. 77.0%) and are similar in s (5.62 vs. 3.71 sec), but the vireo moves at over twice the average speed of the other through the foliage (0.58 vs. 0.26 ft sec).

The relation between feeding behavior and body size in a group of coexisting bird species of similar proportions and general feeding habits is best seen in the Chilean tyrant flycatchers *Muscisaxicola* (Figure 18). As already indicated by the two flycatchers discussed above and as is shown later by the two pairs of tyrants in Chilean and California chaparral (Chapter 5), larger flycatchers are less active, spend more time stationary, and average slower foraging speeds than do smaller flycatchers.

D. Time

It is theoretically possible to treat time as a resource in bird communities inasmuch as bird species might be ranked

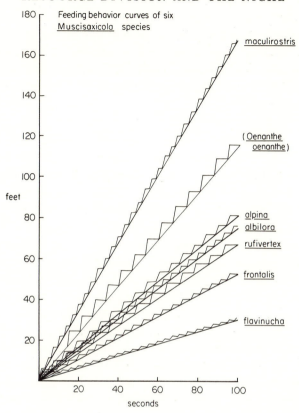

FIGURE 18. Feeding behaviors of six *Muscisaxicola* species from central Chile, and of a northern counterpart, wheatear *Oenanthe*. Body size in *Muscisaxicola* increases from *maculirostris* to *frontalis* and *flavinucha*.

along a time axis in a displacement pattern. But time must be a conceptually different sort of resource from food or habitat, as it is completely exhausted at a uniform rate and yet is perfectly renewable at the same constant rate.

There are several possible ways in which time can be involved in community organization. Birds might use the same resources in the same way, but at a different time of year (for instance, staggered breeding seasons), or at different times of the day, with either a regular alternation of species utilization in a circadian cycle, or simply an asynchronous, opportunistic use of resources throughout the day. Examples of each of these time-dependent displacement patterns can be found, and are discussed next.

1. Breeding Seasons. Bird species that co-occur in a habitat might breed at different times of the year and thereby avoid severe competition with each other. This is not uncommon in some groups of seabirds (Cody, 1973a). Two small petrels, Leach's petrel *Oceanodroma leucorhoa* and the fork-tailed petrel *O. furcata,* breed on Cox Island in the Queen Charlotte group of British Columbia. They are both 7½″ in body size, and appear to be ecologically identical in every way except one: *O. furcata* begins laying in mid-April and *O. leucorhoa* not until mid-June, a difference sufficient to ensure that there are seldom chicks of both species on Cox Island at the same time. British petrels show the same phenomenon, and offset breeding seasons occur in at least one other seabird group, the cormorants *Phalacrocorax.* Most north temperate coasts support two cormorant species, one large and brownish and the other smaller and glossy. The smaller species tends to feed further offshore than the larger, but a much larger contribution to ecological isolation is displacement in their breeding seasons. These are distinct enough on the Olympic Peninsula, Washington state, that the young of the larger species are leaving the breeding ledges when the young of the smaller are hatching, or are just a few days old. The order of the larger and smaller species in the sequence appears unimportant, for the larger of the two English cormorants nests later and the smaller species is the earlier breeder.

A seasonal displacement of breeding seasons in birds with a similar feeding ecology has been reported in the nectar-feeding hummingbirds of California. Anna's hummingbird,

Calypte anna, breeds early in both dry, open chaparral and in wetter canyon sites. It is replaced seasonally by black-chinned hummingbird, *Archilochus alexandri,* in the canyons and by Costa's hummingbird, *C. costae,* in the dry sites (Stiles, 1972). The inference of competition is strengthened by the fact that *costae* breeds earlier in nearby desert habitats which it alone occupies.

Although a great deal of attention has been paid to the timing of breeding in terrestrial species, the chief aim has been the elucidation of the adaptive advantages in a particular species in breeding at a particular time of year. The selective factors involved are the matching of the time when parents will be feeding young to the time when their food supply is most abundant (e.g., Lack, 1968), although age of the parent and the food level when the parent is forming eggs may be also involved (e.g., Perrins, 1970). Within a single trophic group such as insectivorous birds, these selective forces must be similar, for, in almost all temperate communities I have studied, the interspecific displacement in breeding season seems negligible. This is true at least insofar that all the breeding species of the habitat can be found feeding young durifg the same two- or three-week period of the year (although some may have rather longer or shorter breeding seasons than others, and may be displaced somewhat earlier or later). Ricklefs (1966) showed that, on average, any particular species uses 1/1.4 of the total length of time birds breed in its locale and this ratio does not vary between habitats or latitudes. I restricted my own observations on ecological overlap to the peak of the breeding season when all species, scarcely without exception, were feeding young, and thus time of breeding was not counted a factor.

2. Temporal Segregation within a Diurnal Cycle. It appears that in no terrestrial community is any small passerine active at one time of day and not another, while a potentially competing species is active asynchronously with it. This does happen in raptors, as many habitats support both nocturnal and diurnal predators which overlap considerably in diet (e.g., the diurnal *Buteo* hawks and the nocturnal owls such as *Bubo* and *Strix*). Cody and Brown (1969) found that the two most common insectivores of California chaparral are asynchronous in the peaks of singing activity, and interpreted this as the result of competition for "broadcasting time," as songs would be more effective when delivered against a quieter sound background. These species, however, are foraging continuously and even when singing 10 songs a minute they are able to feed actively. Moreover, the peak song activity and interspecific displacement of song periods occurs early in the breeding season when territories are being established, and singing activity is markedly diminished later in the season when the parents are feeding young. Another two chaparral species of the genus *Pipilo* may use habitat patches asynchronously such that individuals do not co-occur in these patches at any given moment of time (see Chapter 6), but again each is feeding continuously somewhere in the habitat, and the behavior does not effect an interspecific displacement onto different resources and reduce competition between them.

In conclusion, time is not a large factor in the organization of terrestrial bird communities, when these are studied over a few weeks at the peak of the breeding season. The broad seasonal shifts in the species composition of communities and the resultant changes in ecological interactions and relations will be discussed as a separate topic (Chapter 4).

III. NICHE THEORY

The chief aim of this monograph is to present the patterns of organization of bird communities such that the selective basis of such patterns is elaborated. A concept that is now in wide use deals just with these same interspecies and community interactions and casts them into a formal structure, niche theory. In order that we can later use the concepts and terminology of niche theory. I will give here a brief review of the development and current tenets of the niche concept.

A. The Evolution of the Niche Concept

The history of the basic ideas of niche theory began, as with most of population biology, with Charles Darwin, who discussed in some detail the possibility of coexistence between ecologically related forms (1859) :

> As the species of the same genus usually have, though by no means invariably, much similarity in habits and constitution, and always in structure, the struggle will generally be more severe between them, if they come into competition with each other, than between the species of distinct genera. We see this in the recent extension over parts of the United States of one species of swallow having caused the decrease of another species.[5] The recent increase of the missel-thrush in parts of Scotland has caused the decrease of the song-thrush. . . . We can dimly see why the competition should be most severe between allied forms, which fill nearly the same place in the economy of nature . . . ."

J. B. Steere perceived thirty-five years later this same association of close taxonomic affinity and separation of range or habitat and suggested how coexistence can nevertheless be achieved (1894) :

> In 17 genera and 74 species each bird genus is represented in the [Philippine] Islands by several species, two or more of which may be found inhabiting the same island; but the species thus found together, with the same generic name,

[5] Perhaps the extension from west to east of the cliff swallow (*Petrochelidon pyrrhonota*) at the expense of the barn swallow (*Hirundo rustica*), the colonization part of which is well documented by, for instance, Baird, Brewer, and Ridgeway (1874).

differ greatly in size, colouring or other characteristics, and belong to different natural sections or subgenera. These sections or subgenera themselves may each be represented in the archipelago by several species; but where this occurs each species is found isolated and separated from all the other species of the subgenus. . . . No two species structurally adapted to the same conditions will occupy the same area.

In 1917 Joseph Grinnell, the pioneering Californian naturalist with such great insights in ecology, coined the term *niche* in reference to the life habits of the California thrasher *Toxostoma redivivum,* but used the concept much earlier (1904) when contrasting two species of chickadees:

> Two species of approximately the same food habits are not likely to remain long evenly balanced in numbers in the same region. One will crowd out the other; the one longest exposed to local conditions and hence best fitted, though ever so slightly, will survive, to the exclusion of any less favored would-be invader, . . .

and later (1917):

> It is, of course, axiomatic that no two species regularly established in a single fauna have precisely the same niche relationships.

But species have all manner of neighbors, some close competitors and some not, and their interactions are potentially independent of taxonomic affiliation. Charles Elton (1927) saw the niche as the species role in a community which provides a set of constraints on ecological activity:

> Animals have all manner of external factors acting upon them—chemicals, physical, and biotic—and the "niche" of an animal means its place in the biotic environment, *its relations to food and enemies.* When an ecologist says "there

50

goes a badger" he should include in his thoughts some definite idea of the animal's place in the community to which it belongs, just as if he had said "there goes the vicar."

The niche of an animal can be defined to a large extent by its size and food habits. . . . The importance of studying niches is partly that it enables us to see how very different animal communities may resemble each other in the essentials of organization.

Competitive exclusion and the niche concept became more strongly associated after the competition equations of Volterra were experimentally tested by G. F. Gause (1934); the so-called exclusion principle is attributed nominally to the work of these two workers, although it did not originate with them. In Gause's words:

. . . as a result of competition two similar species scarcely ever occupy similar niches, but displace each other in such a manner that each takes possession of certain peculiar kinds of food and modes of life in which it has an advantage over its competitor.

Gause was apparently the first to make the connection between natural selection, competition, and the niche—that these concepts are in reality extensively overlapping aspects of the same process (1934):

It seems to us that there is a great future for the Volterra method here, because it enables us not to begin the theory of natural selection by the coefficient of selection but to calculate theoretically the coefficient itself starting from the process of interaction between the two species or mutations.

B. Quantification of the Niche

A quantification of the niche was begun by G. E. Hutchinson (1958), who described species by their ranges along environmental axes important to their biology. If there are k such axes, the niche is a particular region of k-dimensional space, a "hypervolume." This viewpoint has provided a basis for the considerable development of the niche concept in the last few years, particularly by Richard Levins (1968) and Robert MacArthur (1968; 1970). Levins pointed out that the only axes, or *niche dimensions,* we need to consider are those variables which serve to separate species and along which species may be serially arrayed. This restriction alleviates a possible source of bewilderment in the Hutchinson model. Levins has further extended the Volterra competition equations to n coexisting species:

$$dX_i/dt = r_iX_i \frac{(K_i - \Sigma\alpha_{ij}X_j)}{K_i},$$

the growth rate of species i in the presence of competitors j. At equilibrium, the term in the brackets is zero, a condition which can be written in the matrix equation $\mathbf{K} = \mathbf{AX}$, where $\mathbf{K}$ is a vector of carrying capacities, $\mathbf{X}$ a vector of equilibrium densities, and $\mathbf{A}$ a matrix rank n of competition coefficients $[\alpha_{ij}]$. It is then possible to analyze the community matrix (its determinant, eigenvalues, and the mean and variance of $[\alpha_{ij}]$) for information on various community properties, such as its stability, the average packing level of species, and its ecological makeup.

Community equilibrium is only achieved when both the numbers of individuals of all species X and the amount of resources of all the types R utilized by these species are unchanging. Thus for a community equilibrium $dX_i/dt = 0$ and $dR_k/dt = 0$ for all i and k. MacArthur's work points out the precise duality between the resources which support the community (e.g., insect populations) and the species which feed on them (e.g., insectivorous birds). The relations between resources and exploiting species can be represented either as sets of resource curves showing the way each resource is used by similar phenotypes or species, or as sets of species' utilization curves showing to what extent each species uses a set of related resources (Figure 19a and b).

It now becomes possible to relate sets of species that have

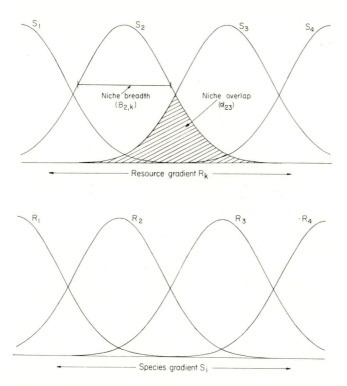

FIGURE 19. Dual representation of niche relations between exploiting species and resources utilized.

assorted themselves into a displacement pattern to niche theory. The resource gradient along which species may show partially nonoverlapping distributions (Figure 19a) is a *niche dimension;* the use of resources by a particular species i is described by a *utilization curve* $U_i(X)$, a picture of its *niche* in this dimension. A measure of the spread of this curve, such as its variance or an information-theoretic value of the diversity of resource use, is a measure of *niche breadth* $B_{i,k}$ of species i on R_k, and the extent to which resource use overlaps between species pairs is shown by the overlap in their utilization curves, or niche overlap $\alpha_{ij,k}$ between species i and j on R_k.

This simple scheme is readily expanded to several dimensions k, such that species niches are represented by clusters of points in k space with a characteristic mean and density distribution. As we are concerned only with those dimensions which are already identified with environmental variables for which competition takes place, as indicated by resultant displacement pattern, three or four niche dimensions are sufficient and manageable. The scheme also lets us think much more sensibly about the "competitive exclusion principle," which is so poorly named and so poorly applied. Competing species do not completely exclude each other, but show considerable overlap in resource use. Moreover, there are some combinations of resource distribution and density which actually favor convergence in utilization curves. Thus the questions we should be asking as a consequence of Gause's results should not center on why the "principle" is being "violated," but rather on how much overlap of resource use is tolerable? How does this overlap vary with resource predictability and density? To what extent are species distributions and abundances predictable from a knowledge of resources types and productivity?

CHAPTER TWO

Niche Breadth

I. SPECIALISTS versus GENERALISTS

A. *Allocation of Energy and Resources*

1. Optimality Models of Resource Use. It is not a tautology
that a jack-of-all-trades should be master of none, nor even
is it obviously correct; yet on an ecologist the phrase settles
as comfortably as a nursery rhyme. Our intuition confirms that
the statement holds, but upon its acceptance depends a great
deal of ecological theorizing.

In the terminology of niche theory, the "jack-of-all-trades
and master of none" concept means either that phenotypes
(species) with broad utilization curves cannot also be phenotypes with high utilization curves, or that phenotypes with
broad utilization curves in one niche dimension cannot have
broad utilization curves in other niche dimensions. Species
with broad niches in a particular dimension have come to be
known as "generalists" in that dimension, whereas species with
narrow utilization curves are termed "specialists." The first
attempts to formalize these notions were the work of R. H.
MacArthur (1965 and ff.) and R. Levins. The Principal of
Allocation of Energy states that there are various competing
energy drains in an individual, and that natural selection will
channel or allocate energy to these various drains such that
the maximum number of offspring is reared to reproductive
age. Thus phenotypes or species that allocate a great deal of
energy to, for example, reproduction cannot simultaneously
maximize adult survival, and phenotypes that minimize their
chances of being taken by a predator cannot also maximize
their efficiency in gathering food. This view led Levins to the

powerful technique of "strategic analysis" (Levins, 1968), a graphical method for the selection of optimal phenotypes for a particular set of environmental circumstances. In particular, the Allocation of Energy principle suggests that the species which is most capable of harvesting one resource will generally be different from the species most capable of harvesting another, especially if the two resources are quite different from each other.

A number of models has been developed with a view to answering the question: How broad will be the diet of a particular species, given a set of resources? The answer will obviously depend on the relative availability of the resources to the species in question, which in turn will depend on the resource densities in a habitat, the ease with which resources can be found and harvested, and the extent to which these rates are changed by the presence of other species interested in the same resources.

Optimal diet breadth can be regarded as a balance between mean search time and mean pursuit time, as the former decreases and the latter increases with diet breadth (MacArthur and Pianka, 1966). Alternatively, diet breadth is determined by the cost of catching and eating an item encountered, with subsequent calorie benefit, versus the likelihood of encountering a more favorable food item after a period of further (time-consuming) search (Emlen, 1966, 1968). This cost/benefit approach has been greatly elaborated by Schoener (1971). Dethier (1954) drew attention to diet restriction in herbivorous insects, and Levins and MacArthur (1969) produced an explanation for the high incidence of monophagy based on plant abundances and insect survival characteristics. As Schoener points out, all of these models are generally similar in their most powerful and interesting predictions, and this seems to be a good reason for restricting further analysis to the most simple of assumptions and to those variables that appear to be common denominators.

2. Resource Difference and Resource Production. Virtually all of the interesting comments to be made on the question "Under what circumstance should a species restrict its feeding to less than the total range of resources?" come from consideration of just two variables. These are a) the difference between the resources, in terms of the fitnesses of the species that use them, and b) the abundance of the resources in the habitat searched. At this level the analysis of specialists versus generalists reduces to components also present in a consideration of optimal habitat or patch selection for the feeding animals, and shows similarity to both the approaches of Levins with strategic analysis and of MacArthur.

To consider a simple example, I return to the two warblers which live in the pine-oak woodlands in the foothills of Arizona mountains. In this vegetation type, the black-throated gray warbler can presumably harvest insects in oak trees much more effectively than it can harvest insects in pine trees, for its occurs in the oak woodland below this mixed vegetation zone but not in the pines above. The reverse applies to Grace's warbler, which in addition to breeding in the pine-oak woodland also occupies coniferous forest above the pine-oak zone. Because this mixed woodland merges into oak woodland below (at least in many sites) and into ponderosa pine forest above, it is of interest to inquire into the use of its different tree species by the two warblers. Suppose the proportion of oaks in the woodland increases smoothly with distance or altitude from $p = 1$ at the bottom of the zone to $p = 0$ at the top of the zone. Thus there is a gradient over which the spatial arrangement of the two resources, oak insects and pine insects, changes in a particular way. The difference between the two resources for the warblers is reflected in their fitnesses in the two tree types. Let the fitness of each species in the preferred tree type be W—for black-throated gray warbler in oaks and for Grace's warbler in pines—and the fitness of each in the nonpreferred tree type be qW, where q is some value between

0 and 1. And finally the resource abundance enters the scheme as a variable. The warblers must spend some time flying from tree to tree in between foraging bouts within trees, and the time spent traveling in this way detracts from fitness, especially if the tree sought is sparsely distributed. A generalist forager will fly a unit distance between trees and, we suppose, spends a proportion k of its time traveling between trees and $(1-k)$ actually feeding. Thus k is related to tree density; it will be high in areas where trees are sparse, and low where trees are dense. The specialist which seeks out a particular tree that occurs with proportion p will expect to fly a distance $1/p$ units between trees; it will therefore spend proportion k/p of its time traveling and only $(1 - k/p)$ feeding. We can now say how the woodland should be used by the two warblers to maximize fitness.

Consider first what just one warbler should be doing if it had the pine-oak woodland all to itself. If the black-throated gray warbler occurs alone in the woodland, its fitness as an oak specialist is $W^{(1-k/p)}$; if it behaves as a generalist and visits both oaks and pines its fitness is

$$W^{(1-k)p} \cdot (qW)^{(1-k)(1-p)}$$

with components which reflect fitness in oaks and pines, respectively. For particular k and q, fitness is a function of p. The oak specialist fitness falls gradually at first but at an increasing rate as p decreases from unity, whereas the oak generalist fitness falls most rapidly at high p and at a decreasing rate with decreasing p (Figure 20). Thus the two curves intersect at some intermediate value of p, a critical value p^*. p^* is given by $(k/(1 - k)\mathrm{ln}W/\mathrm{ln}(1/q)$. At values of p above p^*, the black-throated gray warbler should be an oak specialist and ignore the pines, but at values below p^* it should visit trees as they are encountered and act as a generalist. In sparser woodlands with higher k, the switch from oak specialist to generalist along the gradient should be made sooner, at higher

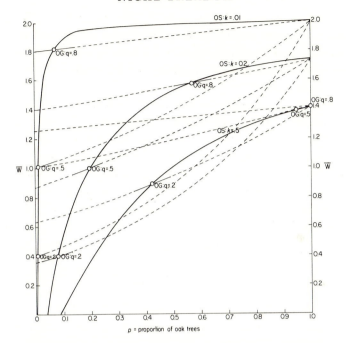

FIGURE 20. Relations between fitness $\overline{W}$ and proportion of oak trees in the habitat of an insectivorous bird which behaves as a generalist (OG) and uses both oaks and pines, or as a specialist (OS) that is restricted to oaks. k is an index of tree spacing, proportional to time spent traveling between trees, and q is the ratio of fitness in less preferred pine trees to fitness in more preferred oak trees. The intersection of the solid specialist curves with the dashed generalist curves determines a switch point p on the habitat gradient where feeding strategies should change.

proportions of oaks, and as relative fitnesses in oaks and pines become more different (q decreases) the switch from specialist to generalist should be delayed until fewer oaks and more pines are encountered. Both lowered resource abundance and increased resource similarity favor a generalist over a specialist strategy.

In this model the general restriction $1 > p > k$ holds, so

59

that p^* exists if and only if $W^{-(k/(1-k))} > q > W^{-(1/(1-k))}$. When q exceeds the left-hand upper limit, a species must be generalist over the whole gradient, and when q is less than the right-hand lower limit, the species must act as a specialist over the whole gradient.

We can now consider the effects of adding another species, the Grace's warbler, which is a pine specialist. The general result can be foreseen, that the switch from oak specialist to generalist will be delayed in the presence of the pine specialist, and vice versa. Thus the competitor, by reducing the value of the nonpreferred resource, acts in a similar way to decreasing q or decreasing k. Suppose the oak generalist contacts the pine specialist. When alone, their fitnesses are

$$OG:\ W^{(1-k)p} \cdot (qW)^{(1-k)(1-p)}$$
$$PS:\ W^{(1-k/(1-p))}.$$

Thus the total fitness derived from pine is

$$(\text{from } OG)(qW)^{(1-k)(1-p)} + (\text{from } PS)W^{(1-k/(1-p))},$$

and this is shared proportionally by the two species when they occur together. Therefore coexistence between the pine specialist and the oak generalist will reduce fitness to

$$OG:\ \frac{q^{2(1-k)(1-p)}\,W^{(1-k)(2-p)}}{W^{(1-k/(1-p))} + q^{(1-k)(1-p)}\,W^{(1-k)(1-p)}}$$

$$\text{and } PS:\ \frac{W^{2(1-k/(1-p))}}{W^{(1-k/(1-p))} + q^{(1-k)(1-p)}\,W^{(1-k)(1-p)}}$$

Parallel expressions describe the way fitnesses are reduced when a pine generalist encounters an oak specialist. When oak and pine generalists co-occur their fitnesses are

$$OG:\ \frac{W^{2(1-k)p}(qW)^{2(1-k)(1-p)}}{(W^{(1-k)p} + (qW)^{(1-k)p})(W^{(1-k)(1-p)} + (qW)^{(1-k)(1-p)})}$$

$$= \frac{q^{2(1-k)(1-p)}\,W^{2(1-k)}}{W^{(1-k)}(1 + q^{(1-k)p})(1 + q^{(1-k)(1-p)})}$$

and likewise for PG with p replaced by $(1-p)$. Fitness, when both species are specialists, is of course unaffected by the pres-

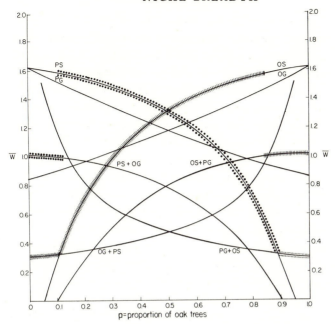

FIGURE 21. Feeding strategies of two coexisting species on habitat gradient, where each species reaches maximal fitness at the opposite end of the gradient. The pine specialist (heavy dots) switches to a generalist feeding mode at p* = 0.89 (light dots), while the oak specialist (vertical hatchings) becomes a generalist (opposite hatchings) when p falls below 0.11.

ence of a second species. The relations between fitness and oak tree proportion are given in Figure 21, for $k = 0.3$, $q = 0.4$, and $W = 2$. On this plot curves represent OS, OG, PS, and PG fitnesses without competition; other curves show fitness under competition between OG and PS (written $OG \times PS$), $OS \times PG$, $PS \times OG$, and $PG \times OS$. The curve for generalists in the presence of the other generalist is not shown, as this combination is never optimal in these circumstances. Whereas the black-throated gray warblers would have switched from oak specialist to oak generalist at $p = 0.32$, in

61

the presence of Grace's warbler the switch is delayed until $p = 0.12$; again the same is true for Grace's warbler if the p's are the proportions of pines in the habitat.

The actual presence or absence of warblers at any point along the gradient is determined by whatever minimal fitness will ensure persistence there. Extremely low fitnesses would be ruled out if the individuals in question did not raise enough young or gain enough experience to offset the risks undertaken. Suppose all fitnesses above some value W_{min} are permissible and worthwhile for the breeding individual. Then the possible combinations of specialist or generalist black-throated gray and Grace's warblers are given in Figure 22, along with the mixes of oak and pine trees at which the changes should occur. For most values of W_{min} the warbler pattern goes from oak specialist to oak specialist-and-pine specialist to pine specialist, but, if low fitnesses are permissible, pine generalist and oak generalist, respectively, are added to the ends of the series to coexist with the other specialist. If only high fitnesses are permissible, neither species may be present in the middle of the gradient, and a hiatus of ranges can result. Perhaps a third species with rather better fitnesses on the equal mixes of oaks and pines could invade successfully. MacArthur (1972) shows by other methods that coexistence of two species X_1 and X_2 on a gradient of resources R_1 and R_2 is influenced by the amount of R_1 and R_2 combined that can be supported at some point on the gradient; the same series of 1–2–1 to 1–0–1 species is predicted as productivity goes from high to low values relative to the harvesting efficiency of the two species. The model just discussed duplicates this result exactly, for when $k = 0.3$, $q = 0.4$, and $W = 2$, the pattern of species coexistence can vary, for different values of W_{min}, from two species coexisting along the length of the gradient through 1–2–1 and 1–1 combinations to a 1–0–1 distribution at high W_{min} (Figure 22).

The variables in this scheme are a) habitat makeup with respect to two resources, oak trees and pine trees (p), b)

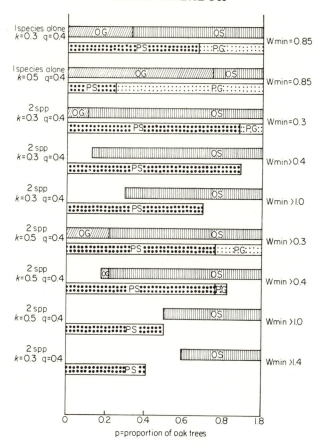

FIGURE 22. Feeding strategies and coexistence patterns in two insectivorous birds with opposite tree preferences on a habitat gradient of pine-oak, for various values of tree spacing (k), relative fitness of the two tree types (q) and W_{min}, the level at which the cost of breeding becomes acceptable. High W_{min} for K-selected populations, low W_{min} for r-selected populations.

abundance of the two resources, measured by the time taken up in traveling between trees (k), c) and the fitnesses of two exploiting species on preferred resource (W) and nonpreferred resource (qW). The first two of these are very easily measured. The second two can most easily be measured by plotting territory size/breeding success in habitats of different types. For instance, territory size of black-throated gray warblers in the oak woodland below the pine-oak zone should be small and may estimate the inverse of W. This species also breeds in piñon pine on California mountain foothills, and its territory size there must more closely reflect qW. Likewise the Grace's warbler can be studied where it occurs without the first species, and perhaps without other serious competition from warblers. In turn, once the expected optimal pattern of species distribution is worked out, it can then be compared to that actually observed. Thus W_{min} can be estimated, which in turn gives an insight into the overall demography of the species. Additional factors might also be looked into. For instance, the productivity of insects by oak and pine trees might differ, and affect the optimal switch points of the gradient (by affecting relative $qW/W = q$). It would also be of interest to incorporate the effect of the third and less common juniper tree into the scheme, perhaps as some sort of buffer resource. Only the black-throated gray of the two warblers has been seen to use the junipers as foraging sites. Preliminary observations at one point on the pine-oak gradient, where p (oaks) = 0.33 and proportion 0.23 of the total number of trees was juniper, Grace's warbler behaved as a specialist with 100% of its time spent in pine trees. The black-throated gray warbler, on the other hand, spent 63% of its time in oaks, and the rest was divided between pines (22%) and junipers (15%); it technically behaved as a generalist ($\chi^2 = 41$, df. = 2, $p < 0.01$).

Qualitative conclusions from the model can be summarized as follows:

a) *Resource abundance.* High resource abundance or high productivity (low k) favors specialists over generalists, and low resource abundance (high k) favors generalists.

b) *Closed canopies*. For foragers in closed canopies k, time spent traveling, can be zero only for generalists (oak generalist: $W^p(qW)^{(1-p)}$. For specialists in closed canopies, canopy diameter d can no longer be assumed, as above, to be small compared to traveling distance, and the lower limit to traveling time proportion becomes $kd(1/p - 1)$. Thus there is to be expected a greater fitness of generalists over specialists closed canopy habitats, particularly where crown size is large.

c) *Resource similarity*. A mix of similar resources (high q) will favor generalists over specialists, and dissimilar resources will support specialists.

d) *Coexistence of exploiting species*. The effect of adding a competitor effectively decreases q, and favors a wider range for specialists. There is a large range of environments over which specialists will coexist, a smaller range where a specialist will coexist with a generalist, but two generalists should not co-occur. Further, when two species are available, a generalist will never occur alone, but always in coexistence with a specialist.

e) Where the premium on potentially low-returns breeding is high, (low W_{min}), extended co-occurrence will be observed. This is equivalent to an r-selection situation, and produces overlapping ranges of species on the environmental gradient. Where high-risk breeding should be avoided (K-selection, W_{min} high) coexistence over the gradient is decreased. Contiguous or abutting ranges, and even hiati in species' combined ranges along the gradient, are to be expected (and are observed in tropical Peru (Terborgh, 1971) and New Guinea (Diamond, 1972)).

II. SELECTION OF NICHE SHAPE

A. Niche Shape as a Measure of Specialization

1. Association of Niche Breadths among Niche Dimensions. In niche theory the arguments about specialists versus generalists become arguments about niche breadth. A food specialist is a species which uses a small range (B_F small) of the food items present in its habitats; a habitat generalist is a species which occupies a broad range of habitats, or B_H is large. But food specialists also have a range of preferred habitats in which they are regularly found, and likewise habitat generalists use a particular distribution of food items. Thus when more than one niche dimension is considered, the specialist-generalist arguments become arguments about niche size and shape; and if niche size remained roughly constant, at least within a broad range of geographic localities, we would be concerned only with niche shape.

Niche size is difficult to measure, for habitat and food selection preferences would have to be measured over the geographic range of a species. A much easier approach to almost the same information is gained by observing the species that occupy some of the habitats in a geographic area and asking: Do these species which occupy the broadest range of habitat types tend to be food specialists or food generalists, and what also of the species with restricted habitat ranges? General information tends to indicate that the habitat specialists are food generalists, and that the food specialists are habitat generalists. This trend, for it is only a trend, is seen in a rough comparison of the finches Emberizidae with the new world warblers Parulidae. Many finches are characteristic of grasslands and brushlands, and are quite habitat specific (see Figure 8). A change of a few inches in mean vegetation height can bring about a species turnover between habitats which are precisely divided on the apparent basis of their structural characteristics. These finches are known to be rather broad in their diets, as many eat both insects and vegetable material in similar proportions. These food generalist habits are shared by other non-finch inhabitants of open habitats, such as larks Alaudidae and blackbirds and meadowlarks Icteridae.

Warblers, on the other hand, appear to have wider habitat

tolerances. Among the most widely distributed species in North America are yellowthroats *Geothypis trichas,* yellow warblers *Dendroica petechia,* and chats *Icteria virens,* which breed from Canada to Mexico and from the Atlantic to the Pacific. Of course, geographic range is no good indication of habitat niche breadth, but one broad-ranging species, orange-crowned warbler *Vermivora celata,* breeds in both deciduous and coniferous forests, from 6′ chaparral to 250′ *Sequoia* forests, and from dry hillsides to damp willow thickets. Other warblers, such as the black-throated gray warbler, Nashville warbler *Vermivora ruficapilla* and yellow warbler, also breed in a variety of habitats, and a third of the 50-odd North American species breed either from coast to coast or from the northern to the southern border of the United States. And warbler species, as MacArthur (1958) and others have shown, may pack six to nine species into some habitats, where they subdivide their insect food supply by behaving as foraging site specialists. It is also worth noting that many instances of ecological replacements, with geographic ranges that abut or overlap marginally, and also cases of interspecific territoriality, are found in the finches, which are the putative habitat specialists and food generalists. Such distributional phenomena are rare in the warblers, and the theory presented above (illustrated, alas, with warblers!) shows that in habitat specialists these coexistence patterns are to be expected, but might be rare in habitat generalists.

It may be possible a priori to suggest relations between niche breadths on two major dimensions, habitat and food. If niche size remains more or less constant, as implied by arguments on ecological saturation, then B_H, habitat niche breadth, and B_F, food niche breadth, should be complementary ($B_H B_F = $ constant, a hyperbolic relationship). But species that occupy an extremely broad range of habitats may be able to do so only if they are prepared to eat a broad range of foods or to feed in a variety of sites provided by the habitat range. And likewise extreme food generalists may find suitable foods or sites in a large range of habitats. Wintering North American finches show such a pattern, as B_H and B_F are inversely related for low B_F but appear to be positively related for high B_F (Figure 23). Ordering five finch species according to B_F simultaneously

	Field sparrow	Slate-colored junco	Song sparrow	White-throated sparrow	Cardinal
Body size	60	68	71	76	80
Bill length	9.4	10.9	11.4	12.5	19.1
Niche size	15.1	23.4	24.5	32.4	48.3

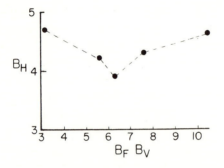

FIGURE 23. Habitat niche breadth B_H versus the product of foraging height niche breadth B_V and diet niche breadth B_F in five wintering finches in eastern North America. Body size order corresponds with niche size order, but habitat niche breadth is minimal in intermediate species. From data in Pulliam and Enders, 1971.

orders the species in increasing niche size (product of niche breadth in three dimensions—habitat, food, and vertical foraging sites), body/bill size and weight, but in decreasing numerical abundance in the sites and in decreasing similarity between the wintering habitat and those selected for breeding. Thus the species which are selected by competition for performance in these habitats year-round are narrow-niched and extremely common there; the species which use the habitat only casually in the winter have the largest niches but are rare there. By extension of these arguments and results, I suggest that continental communities that evolve under severe competition should show a complementarity between niche breadths in different dimensions. In addition, a supplementary relationship may characterize the components of large niches in species that face either a) reduced competitive regimes, as perhaps in species-poor communities (see Chapter 4), or b) extremely rigorous competitive situations such as species-rich wintering grounds (see Chapters 4 and 6). These proposed relations are shown in Figure 24; some quantification will be possible later after discussion of the problems of niche breadth measurement.

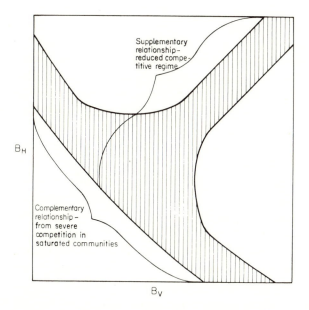

FIGURE 24. Alternative relationships between habitat niche breadth B_H and foraging height niche breadth B_V.

2. *Measurement of Niche Breadth*. Statements about variations in niche breadth are confounded by severe difficulties
in its measurement. Is a species which forages from 2'–4' above
the ground more narrow-niched with respect to vertical feeding distribution than a species which restricts its feeding to
30'–50' in the canopy? This is a problem of scaling, and the
future of much of this area of biology depends on what in-
genious behavioral ecologists can determine about how animal
species view their environments. We already suspect by empirical means that many important variables involved in displacement patterns, such as habitat variables, food sizes, and foraging heights, are discriminated by birds on the basis of relative
rather than absolute magnitudes, and thus the variables are
scaled in terms of some power function. The psychophysicists
are very familiar with this sort of discrimination in humans;
they have produced the Weber-Fechner Law (e.g. Luce,
1959), which says that the just-noticeable difference (jnd)
between two stimuli is proportional to the magnitude of either
of the stimuli. This law holds over a wide range of different
sorts of stimuli in humans, in particular for both visual and
aural signals. If a similar law applies to birds selecting habitats
or foraging heights, it would mean that grass heights which
are measured in a few inches would be discriminated at the
level of an inch or two, but that tall forests whose heights
differed in tens of feet might not be so readily distinguished.

The easiest way to cope with this problem seems to be to
turn it back to the birds themselves, to use the observed
distributions of species along these environmental (stimu-
lus) variables as an indication of how these variables are
perceived. This is the rationale behind the scaling of an
axis of foraging heights in intervals whose absolute magnitude
is small at low foraging height and increasingly larger at
higher foraging heights. Similarly the variables which seem
to govern habitat selection in emberizine finches (Figure 8)
are scaled logarithmically, a move first prompted by observa-

tion of habitat preference in grassland birds (Figure 6). Then, we hope, the intervals along these variables will appear relatively similar in magnitude to the bird species that discriminate among them.

Recently Colwell and Futuyma (1971) have formalized this process; their method is useful as it permits scaling, or distortion, of a resource axis by systematic, algebraic transformations or by any other nonsystematic weighting. The relative "uniqueness" of each "resource state," such as the foraging height intervals just discussed, is measured by looking at the degree to which that resource state is treated as distinct by the species that use it. Then each resource state is weighted according to its distinctness, with the result that the turnover rate of species' utilizations along a series of resource states (such as height intervals) is uniform. Niche breadth is now measured over the new series of resource states such that those treated as distinct have become relatively more common by weighting and those treated as similar have become relatively rarer. This achieves exactly what the subjective quasi-logarithmic scalings used in Chapter 1 did, but now of course the scaling may be performed objectively over a wide range of nonalgebraic functions.

The chief problem in measuring the distinctiveness of resource states is avoiding circularity. If we measure the distinctiveness of resources R_k to bird species by looking at the U_i of all species i using them, we necessarily lose objectivity in subsequent discussions of niche breadths B_i over the resultant modified R_k'. This could be circumvented by measuring the distinctiveness of R_k from some other species utilizations U_j, whose niche breadths we are not interested in measuring. Even if this other data were available, the relevance of the distinctiveness of the R_k with respect to species set j to the distinctiveness of the R_k from the point of view of species set i is questionable. Successional studies, in which resource states and their use by species change over time, can obviously benefit from the Colwell-Futuyma methodology; it was generated from experience in that context. But if the niche breadths of all species U_i using R_k are of interest, and no ecologically related species, or data from them, exist for use as resource distinctiveness indices, the scaling problem still exists. For example, Chilean and California chaparral are structurally similar, and support similar numbers of bird species. The foraging height intervals R_k are logically and potentially the same for these birds in both countries, yet the Chilean species happen to forage over broader vertical ranges than their Californian counterparts. Use of the bird species themselves, at each site, to scale R_k would result in a) fewer distinct resource states, and therefore b) narrower niches in the Chilean birds, whereas the number of resource states must be about the same in the two habitats, and the Chilean species therefore have broader vertical foraging niches.

Although much of the information on how bird species view resource gradients is lacking, some attempt was made in my studies to solve scaling problems. This is most easily accomplished with the intercepts of vertical foraging height, which as already stated was scaled so that the intervals stood an equiprobable chance of including a set percentage of a species foraging time. As all bird species in the community are of interest, and as no other set of species comes close to using the same set of resources, I feel that this scaling cannot be bettered; it at least permits a comparison of B_V distributions between communities. Of course, a best scaling of vegetation height can be expected to change from habitat to habitat, because the differences between the vegetation at, for instance, 6"–2' versus 2'–4', and therefore the expected use of this vegetation by birds, is likely to be greater if some major discontinuity in vegetation density exists between 6" and 4' than if it does not. Thus 2' might be grass top height in a savannah, but might be mid-canopy height in sagebrush. For the sake of maximal objectivity, the same scaling is maintained in all scrub-habitat community studies.

The same fairly objective criterion has been applied to the scaling of habitat variables, along which species distributions determine habitat niche breadth. Such scaling is most easily applied to habitat gradients. In southern California there is a gradient of habitats from the coast inland over the coastal mountains to the interior valleys. The habitats encountered are, from the coast, coastal grassland, coastal sage, south or coastal slope chaparral, north or inland slope chaparral, and oak woodland. Each habitat (about eight distinct types can be recognized) supports some species restricted to that one place in the gradient, but generally species occupy several adjacent habitats in the gradient (Figure 25). As the vegetation varies continually with no ecotones, an objective scaling which also conforms to ecological intuition is one which results by distortion of a combination of the measured habitat variables in normalized species density distributions. Further, the "best" scaling should produce the closest approximation to normalized distributions in all species simultaneously. Thus species which replace each other on the gradient and whose distributions are mutually truncated would not affect the defined "best" scaling of the habitats, as they generally occur in pairs with takeover points at different places on the gradient. Using the simultaneous normalization of species distributions as a criterion for optimal habitat scaling, we again scale the habitat axis logarithmically, using a principle-component combination of vegetation height and vegetation "half-height."

It is less apparent that scaling of habitat variables can assist in the measurement of B_H within a $\pm$ 10-acre study site. In my community studies, the presence or absence of each species in each of the $50' \times 50'$ squares that constitute the site is assessed, and each square is weighted equally. Whereas in the gradient analysis above density is determined by counting the

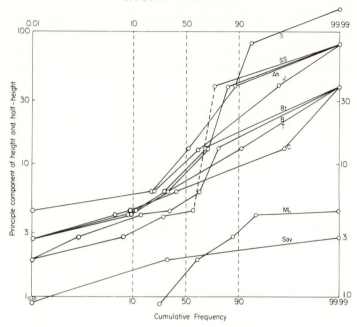

FIGURE 25. Distributions of bird species over a Californian habitat gradient. The habitat gradient is logarithmically scaled to obtain the straightest lines (most normal-like frequency distributions), but even so the distributions are attenuated at the ends. Sav = savannah sparrow, ML = western meadowlark, B = brown towhee, SS = song sparrow, C = California thrasher, T = wrentit, Bt = bushtit, J = scrub jay, Ah = Anna's hummingbird, S = spotted (rufous-sided) towhee.

number of pairs/10 acres, and can be low because either a) territory size is large, or b) only a small part of the 10-acre site is occupied, this within-site measure of habitat niche breadth reflects how patchily a species is distributed at a particular point along the gradient. The contacts between potential competitors are determined by the joint occupancy of the habitat by a pair of species, and can be low only if one or the other of the pair is found on just a small part of the 10-acre site. This unweighted habitat use index is more useful in interspecies overlap measures than in niche breadth computations, but some trends in B_H as the proportion of site occupied are discussed below.

B. Factors Determining Niche Shape

1. Physical Structure of the Environment. The most obvious constraint on niche shape is the physical environment. The disposition of food types over habitats and over foraging sites within habitats will logically determine, to some extent, the selective value of species' restriction to few foods in many foraging sites rather than to many foods at a single site, or something intermediate. This was clearly seen to be the case (above in Section I.A.2) where tree spacing as well as tree similarity determined whether two tree types were treated as one or two resources by warblers. The effect of structure was also obvious in the grassland bird communities (Cody, 1968) where grass height and density limit not only the opportunities for different methods of foraging but also the possibilities for foraging at different heights. The extent of subdivision of grasslands by habitat seems to be a result of whatever can be accomplished by the first two coexistence methods. Figure 26a gives three extreme examples in such grassland communities. Only short grasslands of around 4″ in height permit a variety of feeding behaviors—searching, pursuing, probing, etc.—and only tall grasslands of around 4′ in height permit a stacking of species' vertical foraging ranges one above the other. Grasslands of intermediate height (2½′) permit neither; here the birds subdivide the habitat itself into patches of differing heights that are species-specific, and differ from each other in no other important ways. Thus bill shapes, indices of the foods eaten, are most similar in intermediate grasslands and most different in short grasslands. Niche orientation in grassland birds is summarized in Figure 26b.

Another instance of the effect of habitat structure on niche shape is seen in the tyrant flycatchers. The chief means of

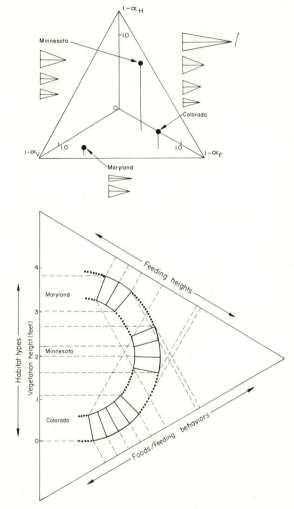

FIGURE 26. Niche relations in North American grassland bird communities, with respect to three resource axes: habitat, foraging height, and food/feeding behavior.

ecological segretation among species of this family, at least in temperate areas, are body size differences (and hence prey size differences) and foraging height differences. Taller woodlands give more opportunity for species to differ in vertical foraging ranges, and thus body size differences between the species that co-occupy those habitats may be de-emphasized. Figure 27 shows an inverse relationship between vegetation

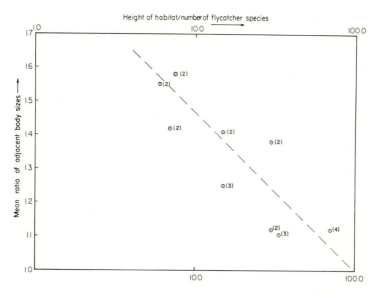

FIGURE 27. Coexistence in North and South American flycatcher species. Body size ratios between species adjacent in the body size series increase as the vertical space available to each species decreases.

height/number of flycatcher species and the average ratio of adjacent body sizes in the series of coexisting species or guilds for one South American and eight North American habitats.

As a final example of structural effects, we consider the North American marine bird species. Some spatial segregation based on habitat differences distinguishes the cormorants (Phalacrocoraciidae), which are chiefly inshore feeders in the northern hemisphere. The two to three species that can be found together on some coasts apparently distinguish between and show differing tendencies to use inshore, estuarine, and fresh-water fishing areas. But the species that fish far offshore, in particular up to at least six coexisting species of auks (Alcidae), have little in the way of cues to distinguish different "habitats" at sea. They nevertheless maintain a spatial segregation, however, based on what may be their only point of reference, the breeding islands; the species form a serial array of foraging distances from the nesting colonies. Among the four species that feed at intermediate distances out to sea, the size of the fish harvested is almost as variable within a species as between species, and there is also a large apparent overlap in the depths to which they dive for food. In the near-absence of food, dive-depth, and habitat differences, then, greater stress is placed on species' segregation by foraging distance from the nest. The two small marine petrels (Hydrobatidae) on the British Columbia coast mentioned above apparently do not have even this coexistence mechanism open to them, and they differ only in the timing of their breeding season (which affords an almost complete separation of activities).

2. *Competitors.* The presence of competing species can be a profound influence on niche shape. The fundamental niche to which a species reverts in the absence of competitors is expected to differ in size, shape, and position from that it normally occupies or realizes; in particular, species are less likely to retain obvious specialization when released from competitor pressure.

To document the effect of competition on niche breadth is to give examples of character displacement (Brown and Wilson, 1956); some have already been given in Chapter 1, and more follow in Chapter 4, where the subject is treated in more detail. The competitive environment of a species can change over a) geographic range or habitats, in particular b) from mainland to island, and also c) over the seasons in one place. It will suffice here to give an example of the first of these, and postpone further discussion.

The illustration above of the effects of vegetation structure of flycatcher niches can be extended. Some widely ranging species, such as the western wood peewee, occur over a range of habitats and come into contact with a variety of other flycatchers in various combinations. Before we ask what effect these competitors have on the peewee's niche, in terms of position (or distribution of vertical foraging heights), we need to know with what freedom different sizes of flycatchers can forage at different heights above the ground. One possibility is that vegetation density affects foraging efficiency. This does seem to be the case, for where three species occur together and are well separated in the vegetation densities they encounter (by dint of foraging at different heights above the ground), the smallest species occurs in the densest vegetation and the largest in the most open vegetation. Figure 28a, which shows flycatcher guilds in one Arizona and three California habitats that each support the western wood peewee, illustrates this effect. However, the relationship between flycatcher size and vegetation density is a relative one; it holds at any one site, but a species may occupy quite different vegetation densities between sites, presumably as a result both of the habitat type and whatever other flycatchers co-occur with it. The peewee is intermediate in size and occupies intermediate vegetation densities, but the absolute densities vary from 0.017 to 0.071. Thus it appears not to be constrained by vegetation density from making a living over a large variety of vegetation types, and therefore over a large variety of heights above the ground.

Peewees are constrained, however, by the presence of other

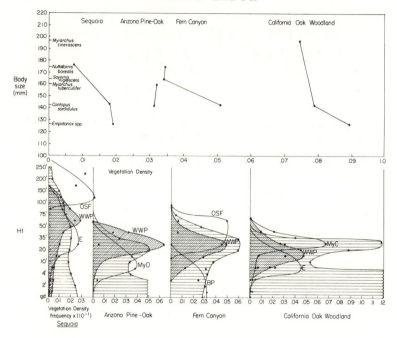

FIGURE 28. Relations between body sizes and foraging height distribution in four flycatcher guilds. Smaller species are always in *relatively* denser vegetation, but not associated with *absolute* vegetation density.

flycatchers. They are restricted to intermediate heights in *Sequoia* forest, mixed riparian woodland (Fern Canyon), and oak woodland, even though the mean foraging height varies from 15′ to about 70′. Only in Arizona pine-oak does no larger flycatcher occur above peewees, and only here do they forage up to the top of the foliage. Here both flycatchers are similar in size and are virtually identical in the mean vegetation densities they encounter. In fact there is a larger flycatcher in the Arizona pine-oak, Wied's crested flycatcher *Myiarchus tyrannulus,* 205 mm, but its occurrence in the pine-oak is sporadic.

3. Resource Predictability and Seasonality. A third factor that might be expected to affect niche breadth is the patterning of resource abundances over time. This patterning has two aspects: a) predictability, which describes the accuracy with which the level of a particular resource can be predicted for some future point in time, knowing the recent levels of the resource, and b) seasonality, the extent to which resources replace each other in level and type over time. Resource predictability could be an important selective force both between seasons and within a season, whereas the resource seasonality factor is strictly a within-season effect.

Both resource predictability and seasonality can be related to niche breadth by simple, logical arguments. The production curve of a resource span can be given confidence limits for a certain time period. If a part of the resource span is characterized by a mean level with high predictability (confidence limits close to that mean), then a species may depend on those resources, can specialize upon them, and afford to ignore other neighboring resources in the span. Because the Principle of Allocation of Energy shows that more overall efficiency is lost the greater the span of resources used, species will evolve utilization curves that in a sense are only as broad as they need be to avoid extinction (as Richard Lewontin once put it). Consequently niche breadths are expected to evolve as a balance between the rewards of adaptive specialization on few resources and the rewards in terms of long-term persistence of a broader resource utilization; thus higher resource predictability should allow smaller niche breadths. On the other hand, high turnover or seasonality of resources should favor broad niches. Phenotypes must now possess a morphology that allows them to exploit one set of resources during one season and another set at a different time of the year. Thus their morphology and/or behavior must be a compromise between (at least) two different selective optima, a compromise that may affect their resource use in one season in the direction

of broader utilization curves. In summary, an optimal phenotype is one that distributes its time and energy optimally
among competing demands, and these demands include, for
instance, foraging in various ways on different parts of various resource gradients. A realized optimal distribution is subject to the constraints of resource variation, and departs from
a nonattainable but higher optimum as this variation (with
both predictability and seasonality aspects) increases.

As already discussed, competitors can affect resource levels
and, because of seasonal turnovers and changes in numbers
of competitors, can affect also the predictability and seasonality of resources. But climate can affect niche breadth directly
by way of its effects on resources, and the community studies
can be used to evaluate this. Ideally we should define steps
in time t_1, t_2, t_r, t_s through a season and on into following
seasons and years (up to at least the end of the expected lifespan of the species concerned) and levels of a resource 1, 2,
m; then there are $s(s-1)/2$ matrices of rank m, one for
each pair of t_r, whose elements are the transition probabilities $[p_{ij}]$ of changing resource level from i to j between the
two time steps. From this information, optimal strategies could
be derived, and, for a complete understanding of the putative
optimal strategies that selection has produced, such a process
might be necessary. But we shall make do with something
much simpler, and I have used the eight climatic variables
described in Table 1. These include five that may measure
some aspect of predictability and three that measure seasonality for both temperature and precipitation. These variables
were used as independent variables in step-wise regression
analyses with niche breadth measures as the dependent variables; their values at each community study site are given
in Appendix A.

Niche breadth measures are available for habitat use
(B_H = proportion of study area occupied by a species) and
for vertical foraging distribution ($B_V = e^H$ where $H =$

TABLE 1. Climatic variables used as estimators of resource variation

<table>
<tr><td rowspan="5">Predictability</td><td>

1. Rainfall variance R_V. Variance of monthly precipitation for each of the breeding season months May, June and July over 20-year period, averaged over the three months' values.
2. Temperature variance T_V. Same as in 1, except substitute monthly mean temperature for monthly precipitation total.
3. Bad weather probability P. The probability that, during the three breeding season months, a species may encounter monthly precipitation totals > 150% or < 50% of the long-term average, or monthly mean temperatures > 105% or < 95% of the long-term averages.
4. Rainfall autocorrelation $R(Y - Y)$. Autocorrelation value derived from time series analysis of 20 years of monthly rainfall totals using 12 month shift.
5. Temperature autocorrelation $T(Y - Y)$. Same as 4, except substitute monthly mean temperature for monthly precipitation totals.

</td></tr>
</table>

6. Rainfall seasonality $R(M - M)$. The mean deviation of monthly rainfall totals from $\frac{1}{12}$ of the annual rainfall total, using 20 years of data as before.
7. Temperature seasonality $T(M - M)$. The mean deviation of monthly mean temperature from the annual mean temperature, in °F, using 20 years of data to find annual and monthly means.
8. Frost-free days (FFD). The mean number of frost-free days a year, 20-year average.

Predictability

Seasonality

$\Sigma_i - p_i ln p_i$ and the p_i are the proportions of the time spent foraging in vertically stacked layers). For regressions involving B_V, the results are clear even if negative: there is no significant relationship between B_V and any of the climatic variables. Using six climate variables (variables 4 and 5 were omitted as they consistently failed to be preferentially included into step-wise regression), only 8% of the variance in B_V is accounted for ($n = 101$ species in eight North American study sites, as insufficient climate information exists for the Chilean sites for their inclusion; regression d.f. = 6, error d.f. $= 94$, $F = 1.33$, $p = 0.25$). Regressions involving B_H were only slightly more encouraging. Using the same six climate variables, we account for 15% of the variation in B_H (a statistically significant if unimpressive figure; $F_{6,94} = 2.96$, $p = 0.02$). The three predictability variables 1–3 alone can do almost as well (15%; $F_{3,97} = 5.57$, $p = 0.002$), whereas the three seasonality measures accomplish considerably less in variance reduction (11%; $F_{3,97} = 4.35$, $p = 0.007$). Figure 29

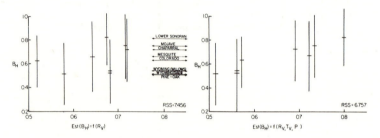

Figure 29. Habitat niche breadth B_H increases with climatic predictability. B_H is estimated with a single regression variable on the left, but is better predicted using a regression of three climate variables (right).

plots residuals at two stages of the regression analysis. I conclude that B_V is not related to climatic (and hence resource) variation, at least within the range of climates tested in North America, but that there is a slight trend for species to occupy only certain, more limited parts of habitats as climatic predictability decreases. Note that this effect is opposed to the finding of MacArthur et al. (1966), who contrasted habitat use in tropical and temperate birds. My results might be due to greater variation in the survival of birds in low-predictability areas, resulting in more patchy habitat occupancy, but additional evidence on niche overlap (below) will support its attribution to a more precise and restrictive habitat selection.

The possibility that B_H values are affected by the amount of habitat variation included by chance in the study sites can also be eliminated. Each study site is divided into $50' \times 50'$ squares, each of which possesses certain structural characteristics; thus the study site is characterized by the vegetation structure of the average $50' \times 50'$ quadrat. The mean overlap in the structural characteristics between this average quadrat and each other quadrat of similar size was calculated; these values of within-site horizontal habitat diversity HHD are given in Table 2. B_H and FFD are uncorrelated ($r = 0.085$), and the latter can account for only 0.73% of the variation of the former.

It remains to investigate the possibility that B_F is related

TABLE 2. Within-site horizontal habitat diversity for community study sites, calculated as the mean overlap in structural characteristics of each $50' \times 50'$ quadrat with the site average $50' \times 50'$ quadrat.

Community study site	Structural characteristics measured in quadrat	Number of categories of quadrat variables	Number of quadrats per site	Mean overlap between quadrats	S.D. of overlap
Teton Willows	Bush ht. proportions in $50' \times 50'$ quadrats	8	98	0.536	0.131
Teton Sagebrush	Proportions of bushes in various height categories per $50' \times 50'$ quadrat	7	221	0.819	0.089
Colorado Saltbrush	As in Teton Sagebrush	7	69	0.360	0.106
Mohave Desert	As in Teton Sagebrush	8	155	0.673	0.123
California Chaparral	As in Teton Willows	9	129	0.354	0.141
Lower Sonoran Desert	As in Teton Sagebrush	8	198	0.426	0.155
Arizona Mesquite	As in Teton Sagebrush	8	215	0.610	0.169
Arizona Pine-Oak	As in Teton Sagebrush	7	112	0.487	0.170

to either resource predictability or seasonality, but this can be done only indirectly. I chose three species properties that are expected to be related to feeding ecology: body size BS, the ratio of bill depth to bill length d/l, and the mean foraging speed v. Each community shows a range of values in each of these variables. Is the range of values found in the community related to resource predictability? To test this, I calculated regression equations with each of the above variables in turn as a dependent variable and the three predictability estimates T_V, R_V, and P together as independent variables. To take body size first, temperature variance T_V was the most significant independent variable and was positively correlated with the mean body size in the community. For each community there is a distribution of residuals from the regression, and the standard deviation of these residuals is strongly correlated with the body size estimated from the climatic data $(r = 0.886;\ p = 0.005)$. Thus body size tends to be both higher and more variable in unpredictable areas. Foraging speed v was most closely associated with P, the probability of encountering unusual weather; unpredictable climates favor low foraging speeds. More significantly, predictable habitats are associated with a large variety of foraging speeds, but the community variance in species' mean foraging speed is low in unpredictable areas $(r = 0.685,\ p = 0.06)$. Finally the ratio of bill depth to length, also most sensitive to P, shows a tendency to increase with climatic uncertainty. This indicates that the proportion of seed eaters over insect eaters increases in areas of climatic uncertainty. The variance of d/l, however, shows no relation to d/l estimated from the regression on climate variables $(r = 0.176,$ not significant$)$.

In summary, the following trends in niche breadths are apparent. In areas of climatic and hence resource uncertainty, species tend to be restricted to specific and structurally dissimilar parts of the habitat and are of larger but more various body sizes. They appear to be more similar in diet, and forage in similar ways. In contrast, species in predictable areas tend to occupy all of the study area (a greater range of habitat variability), and, although they are all of more similar, smaller body sizes, they apparently include more insectivores and employ a greater variety of foraging methods. The similarity in body size is most likely related to the similarity in habitat which they encounter in common, the low value of body size is related to the low temperature variance (or perhaps higher mean temperatures) encountered, and the variety of feeding methods is necessary to allow coexistence among species of similar sizes in the same habitat. This picture will be completed in the following chapter on variation in niche overlaps.

CHAPTER THREE

Niche Overlap

I. MEASUREMENT AND REPRESENTATION OF NICHE OVERLAP

A. *Two Sorts of Niche Overlap*

Ideally niche overlap is measured from species' utilization curves over resource gradients, the distributions of probabilities that a species takes a unit of resource from a particular place on the resource gradient. Then niche overlap is the probability that a pair of species is taking the same resources, and can be measured by MacArthur's formula

$$\alpha_{uv} = \frac{\int u(x)v(x)}{\int u^2(x)}$$

which is the area common to the utilization curves of the two species u and v on the resource gradient x. But only one of the three main niche dimensions discussed in Chapter 1—vertical foraging range—can be, and is, used to obtain a niche overlap value in this way. For habitat overlap I obtain symmetrical overlap values from the ratio of the number of $50' \times 50'$ quadrats two species hold in common, p_{12}, to the geometric mean of the number of such quadrats each species holds over the whole study area,

$$\alpha_{H,12} = p_{12}/((p_{11} + p_{12})(p_{22} + p_{12}))^{1/2},$$

where p_{11} and p_{22} are the numbers of quadrats of each species not occupied by the other. This yields overlaps between zero and one, and sidesteps niche breadth issues. α_F is also conventionalized to give values between zero and one; thus in only one of the three components of niche overlap is it obvious a priori what measure is preferable.

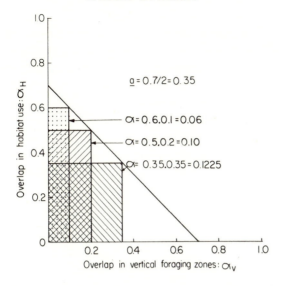

FIGURE 30. With a constant value of summation alpha $\mathbf{a}$, $= (\alpha_H + \alpha_V)/2$, product alpha $\boldsymbol{\alpha}$ varies according to the shape of niche overlap. $\boldsymbol{\alpha} = \alpha_H \cdot \alpha_V$.

The next problem is to combine the three components, which in practice is even more formidable. Earlier, I reasoned (Cody, 1968a) that species may or may not differ in each niche dimension separately, and that coexistence might be achieved by species pairs exceeding some threshold of minimum ecological difference. This led to an average or "summation alpha," represented by $\mathbf{a} = \Sigma^k(\alpha_R)/k = (\alpha_H + \alpha_V + \alpha_F)/3$ in this case. But it is equally reasonable to visualize overall niche overlap as the product of its various components, for if there are probability density functions for each species on each niche axis, then the probability of their joint use of resources is just the product of their separate probabilities. This leads to a "product alpha," $\boldsymbol{\alpha} = \Pi^k(\alpha_R) = \alpha_H \cdot \alpha_V \cdot \alpha_F$.

Robert May (MS) has pointed out that there is no infallible way of estimating multidimensional alpha from the projections of niches separately on the niche axes. For resource axes that are quite independent, product alpha gives the best estimate, but when niche dimensions are nonindependent, summation alpha is the more accurate estimator. Without knowledge of the interdependence of axes we are left with only empirical tests of the two forms. Each is in some sense an extreme form, each correct in certain particular circumstances and each akin to a bound for the true value of niche overlap, which will generally lie between them. The two measures are of course related as $\mathbf{a}^3 \geq \alpha \geq 3\mathbf{a} - 2$ (Figure 31), and in general for n

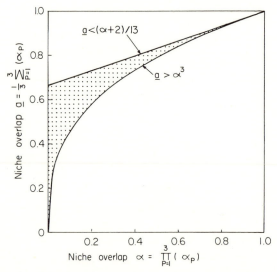

FIGURE 31. The relation between two niche overlap measures. **a** is an upper bound of true niche overlap.

niche dimensions $\mathbf{a}^n \geq \alpha \geq n\mathbf{a} - (n-1)$. The practical way in which the measures diverge is that $\alpha = 0$ if any single component is zero, but $\mathbf{a} = 0$ only if all components are zero. I will show below that **a** behaves in a generally more orderly and predictable way with varying environments than does α, indicating considerable interdependence of niche dimensions. A further point may be made about the difference between summation and product alphas. Species may be competitively excluded from a section of a resource gradient by competitors with which they now show no overlap. But, following the removal of such competitors, the restricted species may well be able to expand its resource utilization to the sections of the resource gradient occupied by the missing species. Thus a zero measure of the previous overlap will not accurately predict niche expansion and density compensation in environments with fewer competitors. Thus a was previously termed "expansion alpha" (Yeaton and Cody, 1973), and was judged empirically as well as theoretically to predict more accurately new equilibrium densities of species on islands with a reduced competitive regime.

B. The Community Matrix

The two forms of niche overlap just discussed are alternative estimates of the competition coefficient α_{ij} in the Lotka-Volterra competition equations. There, α_{ij} is a weighting coefficient, such that the numbers X_j of the competitor are cast as equivalent numbers of X_i. Clearly, as the two species approach identity, the weighting or competition coefficient approaches unity; here it is estimated by niche overlap measures that likewise approach unity as the two species in question approach identity in their apparent use of resources. In competition by bird species for seed and insect resources, α_{ij} is most likely restricted to the range 0–1; only in interference competition or competition between species of very different sizes or efficiencies, two conditions ruled out in passerine bird communities, would α_{ij} take higher positive values.

For each community I have studied (Appendix A), niche overlaps can be expressed in either summation or product form and assembled in the community matrix $\mathbf{A} = [\alpha_{ij}]$. The diagonal elements α_{ii} are unity; the matrices are symmetrical, as assessment of $\alpha_{ij} \neq \alpha_{ji}$ involves severe problems of niche breadth measurement. Figures 32 and 33a give the community matrices for the Wyoming willows and Arizona-pine-oak-juniper sites, and the remainder of these data are relegated to Appendix B.

C. Ecological Relations within Communities

1. Dendrograms of niche overlaps. Perhaps the clearest picture of the ecological relationships within communities is obtained from dendrograms ("tree-pictures"). Dendrograms are usually used to represent genealogies or taxonomic affinities within a set of related species, but here we represent interspecific affinity by the degree of their ecological overlap (as, for example, did Goodall, 1970). Sokal and Sneath (1963) discuss the techniques and limitations of drawing up dendrograms. We begin by combining the two species with the highest degree of niche overlap (A and B), and substitute in the community matrix a single "species" AB whose ecological overlaps with the remaining species is, by convention, the average overlap with each of A and B $\left(\alpha_{C,AB} = \dfrac{\alpha_{CA} + \alpha_{CB}}{2} \right)$.

This is repeated with the highest niche overlap figure in the new matrix (of rank reduced by 1), and so on until a single overlap remains. In the community matrix of the Wyoming willows, Figure 32, the highest overlap value is 0.829, between species 3, yellowthroat, and species 7, Lincoln's sparrow. Thus these two are combined into unit A. Niche overlap between unit A and species 1 is $(\alpha_{13} + \alpha_{17})/2 = (0.594 + 0.672)/2 = 0.633$. Columns and rows 3 and 7 are replaced with a new row of overlaps with unit A. Next, species 10, white-crowned sparrow, joins unit A at $\alpha_{A,10} = 0.755$, to form unit A'. Then the species pair 8 and 11, song sparrow and fox sparrow, with overlap 0.695, combine to form unit B, resulting in a matrix

NICHE OVERLAP

```
          1     2     3     4     5     6     7      8     9    10    11    12
 1    1.000 0.327 0.594 0.478 0.513 0.359 0.672 0.324 0.276 0.618 0.403 0.285    1
 2          1.000 0.473 0.601 0.450 0.564 0.433 0.229 0.418 0.418 0.328 0.328    2
 3                1.000 0.681 0.619 0.401 0.829 0.424 0.470 0.765 0.511 0.337    3
 4                      1.000 0.537 0.530 0.632 0.390 0.431 0.582 0.446 0.353    4
 5                            1.000 0.489 0.689 0.426 0.311 0.612 0.451 0.338    5
 6                                  1.000 0.400 0.310 0.346 0.404 0.299 0.259    6
 7                                        1.000 0.529 0.420 0.745 0.623 0.377    7
 8                                              1.000 0.244 0.417 0.695 0.542    8
 9                                                    1.000 0.388 0.356 0.232    9
10                                                          1.000 0.496 0.366   10
11                                                                1.000 0.439   11
12                                                                      1.000   12
```

```
(3+7)
 = A   0.633 0.453       0.657 0.654 0.401       0.477 0.445 0.755 0.567 0.357   A
(A+10)
 = A'  0.626 0.436       0.620 0.633 0.403       0.447 0.417       0.532 0.361   A'
(8+11)
 = B   0.364 0.279       0.418 0.438 0.305             0.300             0.491 0.490   B
```

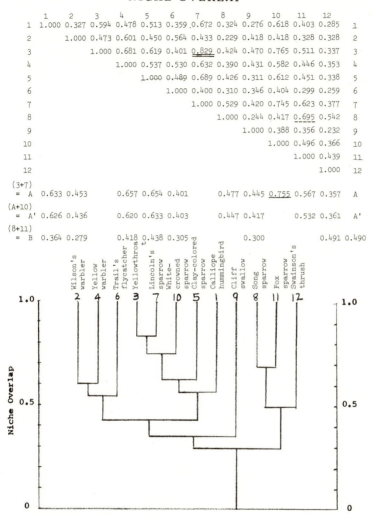

FIGURE 32. Community matrix and community dendrogram for Wyoming willows community.

	1	2	3	4	5	6	7	8	9	10	11	12	13	14	15	16	17	18	19	20
1	1.000	0.403	0.366	0.413	0.432	0.391	0.515	0.381	0.285	0.331	0.555	0.279	0.303	0.483	0.362	0.283	0.302	0.249	0.303	0.297
2		1.000	0.581	0.534	0.586	0.417	0.543	0.522	0.474	0.366	0.463	0.396	0.249	0.526	0.559	0.253	0.505	0.468	0.304	0.145
3			1.000	0.717	0.702	0.481	0.414	0.602	0.592	0.572	0.478	0.428	0.450	0.655	0.650	0.414	0.595	0.450	0.330	0.231
4				1.000	0.860	0.411	0.423	0.509	0.525	0.543	0.494	0.324	0.395	0.597	0.479	0.389	0.485	0.278	0.340	0.154
5					1.000	0.431	0.542	0.630	0.517	0.518	0.571	0.436	0.412	0.601	0.591	0.392	0.524	0.174	0.241	0.380
6						1.000	0.203	0.586	0.520	0.385	0.365	0.490	0.339	0.384	0.499	0.294	0.555	0.407	0.134	0.144
7							1.000	0.349	0.344	0.392	0.581	0.469	0.298	0.327	0.491	0.250	0.359	0.354	0.339	0.187
8								1.000	0.677	0.535	0.530	0.627	0.625	0.473	0.619	0.439	0.662	0.508	0.315	0.371
9									1.000	0.400	0.482	0.404	0.480	0.154	0.584	0.268	0.683	0.590	0.393	0.373
10										1.000	0.419	0.384	0.357	0.492	0.559	0.475	0.539	0.404	0.649	0.667
11											1.000	0.414	0.290	0.572	0.553	0.328	0.392	0.343	0.347	0.407
12												1.000	0.307	0.458	0.646	0.268	0.593	0.520	0.243	0.232
13													1.000	0.420	0.429	0.436	0.466	0.431	0.239	0.183
14														1.000	0.579	0.392	0.533	0.480	0.423	0.332
15															1.000	0.344	0.687	0.582	0.444	0.283
16																1.000	0.354	0.286	0.329	0.216
17																	1.000	0.468	0.414	0.217
18																		1.000	0.468	0.315
19																			1.000	0.655
20																				1.000

FIGURE 33a. Arizona Pine-Oak community matrix.

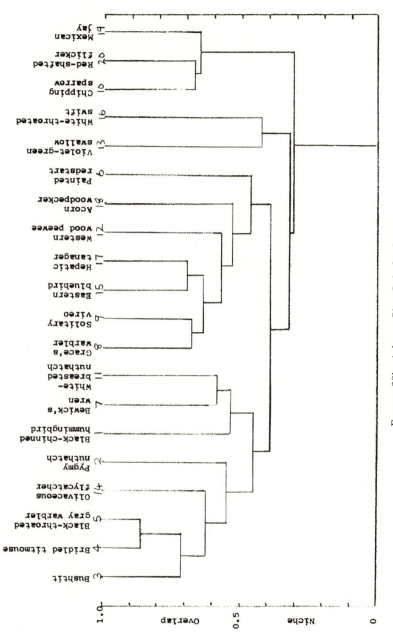

FIGURE 33b. Arizona Pine-Oak dendrogram.

of 9×9 of overlaps between species 1, 2, 4, 5, 6, 8, 9, 11, 12, A', and B. The eventual result is an assorting of species as shown in Figure 32. Figure 33b shows the dendrogram of ecological relations for Arizona pine-oak, and those for the remaining communities are given in Appendix B.

2. Cluster Groups: Specialists and Guilds. Ecological dendrograms tell at a glance which species are most closely associated with which others. They show how the community may be broken down into smaller clusters or groups within which competition is likely to be more intense than between members of different groups. These groups of ecologically similar species have come to be known as "guilds." The dendrograms also show which species are ecologically isolated from the other community members and stand alone as specialists.

In the willows (Figure 32), three broad clusters are distinguishable: a group of three species (left) of insectivores that feed rapidly and high; a group of five species (center) that feed lower in the foilage and more slowly; and three species (right) that feed almost exclusively on the ground, with slow, searching movements. These guilds consist of core species, such as Wilson's warbler, yellowthroat, and song sparrow, which initiate the cluster or join it early, and fringe species, that differ considerably from the others in the group and tag onto the outside as specialists. Examples of the latter are cliff swallow, Calliope hummingbird, and Swainson's thrush.

Core species generally show large mean niche overlaps with the rest of the community (measured by the average of the appropriate row or column elements in the symmetric community matrix), whereas fringe species show low community overlaps. These figures are given in Table 3; community overlaps for each member species i are computed as $\Sigma_j^{n-1} a_{ij}/n - 1$ exclusive of the diagonal term a_{ii}. Thus yellowthroat, Lincoln's sparrow, yellow warbler, and white-crowned sparrow are largest in community overlaps at 0.555, 0.553, 0.515, and

TABLE 3. Row/column means of the symmetrical community matrix a measure, of the competitive involvement of each species with the other members of the community; Wyoming willows and Arizona pine-oak sites.

Willows			Pine-Oak		
	Community overlap	Rank		Community overlap	Rank
1. Calliope hummingbird	0.441	6	1. Black-chinned hummingbird	0.365	4
2. Wilson's warbler	0.415	5	2. Pygmy nuthatch	0.437	10
3. Yellowthroat	0.555	12	3. Bushtit	0.511	18
4. Yellow warbler	0.515	9	4. Bridled titmouse	0.467	12
5. Clay-colored sparrow	0.494	8	5. Black-throated gray warbler	0.502	17
6. Trail's flycatcher	0.396	3	6. Painted redstart	0.391	7
7. Lincoln's sparrow	0.553	11	7. Bewick's wren	0.388	6
8. Song sparrow	0.412	4	8. Grace's warbler	0.524	20
9. Cliff swallow	0.354	2	9. Solitary vireo	0.476	14
10. White-crowned sparrow	0.528	10	10. Chipping sparrow	0.473	13
11. Fox sparrow	0.459	7	11. White-breasted nuthatch	0.452	11
12. Swainson's thrush	0.351	1	12. Western wood peewee	0.417	9
			13. Violet-green swallow	0.374	5
			14. Olivaceous flycatcher	0.483	15
Mean	0.458		15. Eastern bluebird	0.523	19
			16. White-throated swift	0.337	2
			17. Hepatic tanager	0.491	16
			18. Acorn woodpecker	0.409	8
			19. Mexican jay	0.364	3
			20. Red-shafted flicker	0.305	1
			Mean	0.434	

0.528, respectively, while cliff swallow, Trail's flycatcher and Swainson's thrush are lowest at 0.354, 0.396, and 0.351. Calliope hummingbird is somewhat less ecologically isolated, but easily qualifies as a specialist at 0.441, as it is well below the community-wide average overlap of 0.458.

Clusters or guilds can form for various reasons. The perhaps surprising initial combination of yellowthroat and Lincoln's sparrow comes about because the two have similar ecologies all round. They show almost identical habitat preference $(\alpha_{H;3,7} = 0.922$, versus a community-wide average of 0.524), overlap greatly in foraging heights $(\alpha_{V;3,7} = 0.830$, versus a community average of 0.254), are very similar in feeding behavior $(\alpha_{F,Be;3,7} = 0.853$ versus a community average of 0.719), and are more similar than average in bill morphology

$(\alpha_{F,Bi;3,7} = 0.863$, community average $0.837)$. The white-crowned sparrow joins the guild because it uses the same habitat as the yellowthroat $(\alpha_{H;3,10} = 0.824)$ although rather different in other characteristics, and because its behavior and morphology are very similar to those of the Lincoln's sparrow $(\alpha_{F,Be;7,10} = 0.928;\ \alpha_{F,Bi;7,10} = 0.946)$, although again it is rather different in the remaining dimensions. Song and fox sparrows are, again, generally similar all-round, but particularly in foraging heights $(\alpha_{V;8,11} = 0.740)$, feeding behavior $(\alpha_{F,Be;8,11} = 0.848)$, and bills $(\alpha_{F,Bi;8,11} = 0.900)$; Wilson's warbler and yellow warbler, while quite different in habitat preference $(\alpha_{H;2,4} = 0.480)$ and behavior $(\alpha_{F,Be;2,4} = 0.641)$, forage at similar heights $(\alpha_{V;2,4} = 0.720)$ and possess similar bills $(\alpha_{F,Bi;2,4} = 0.940)$.

In the Arizona pine-oak, the community is rather less easily broken down into discrete guilds. Species are more equitably associated with each other, and show a reduced tendency to form exclusive, tight clusters. This difference is illustrated by the distribution of species community overlaps in Table 3; the variance among the row averages is reduced in the pine-oak over the willows.

The pine-oak supports two broad guilds of chiefly foliage insectivores, one of eight species to the left and one of seven in the center. In addition there is a pair of aerial insectivores and a trio of chiefly ground-searching species to the right. The foliage-insectivore guild to the left centers on black-throated gray warbler and bushtit, lower and faster species, while the guild in the center includes higher and slower insectivores. Trunk searchers are included in the former cluster. Within each of these two guilds, species owe their relative positioning to similarity in feeding behavior, while the main differences between guilds are in foraging height.

3. Ecological Counterparts. Inasmuch as the habitat structure of the two sites is similar, we expect the two bird communities found there to be similarly organized. A comparison of the community dendrograms can show whether or not this is true. Thus, in addition to the easily quantified comparisons of species number and relative abundance, a further and higher order comparison is made possible. This subject is treated in detail in Chapter 5, but some comments on ecological counterparts in the eight North American and three Chilean study sites (Figures 32, 33, 59, 60, Appendix B) are made now.

The clustering process results in some consistent relations among communities. Usually two major groupings are formed, one of foliage feeders and one of ground feeders (sites 1, 3, 4, 5, 10, 11,). Two additional sites also show this, but include trunk feeders respectively with ground feeders (site 6) or foliage feeders (site 8). Most frequently, the aerial feeders such as swallows, swifts, and nighthawks are classified as specialists ecologically distinct from either of these first two major categories (sites 1, 4, 5, 6, 7, 8, 10). Hummingbirds, present at 8 of the 11 sites, are usually classified as specialists on the fringe of foliage feeders (sites 1, 7, 10) or further within that group (sites 4, 5, 6).

Within the large cluster of foliage feeders, further breakdowns can be made in common at several sites into generally smaller and more rapid feeders versus larger and slower feeders. Species within each of these guilds also have in common their foraging height distributions, but the faster species may be the lower of the two guilds (sites 4, 8) or the higher (sites 1, 5).

Such comparisons between community organization patterns could proceed to the point of attempted one-for-one matching of individual species, but this is only worthwhile when the habitats are structurally well matched and only interesting when the resident faunas are taxonomically distinct (see Chapter 5). Most of the obvious intersite community organization differences are easily associated with differences in habitat structure, such as the large number of trunk feeders in the only woodland site (8), and the larger numbers of ground feeders in the drier and more open desert and desert-like sites (2, 3, 4, 6, 7, 9, 11).

II. WITHIN-COMMUNITY VARIATION IN NICHE OVERLAP

A. Factors Affecting Niche Overlap

1. Body Size. We next analyze patterns of variation in niche overlap: in this section (II), variation in the pair-wise overlaps within communities is considered; in Section III we look at patterns in mean overlaps shown by the range of communities.

We discussed earlier a common coexistence mechanism in guilds or between congeners, size differences, which in turn was linked to diet difference. It is expected, then, that species differing in size will show reduced niche overlap **a**, as components of **a** are factors that should strongly affect diet. Table 4 shows that this is so. Niche overlap was measured between

TABLE 4. Variation in niche overlap components with body size. 1st order pairs are species with body size ratios less than 1.3, 2nd order pairs have body size ratios between 1.3 and 1.69, and 3rd order pairs are species with body size ratios in excess of 1.69. Each figure represents an average of 11 study communities.

	All pairs	1st order pairs	2nd order pairs	3rd order pairs	Significance of differences between the means		
					1–2	2–3	1–3
n	990	378	283	329			
a	0.476	0.509	0.485	0.430	0.02	<0.001	<0.001
α_h	0.634	0.639	0.653	0.613	>0.05	0.04	>0.05
α_V	0.321	0.347	0.291	0.212	0.03	<0.001	<0.001
α_F	0.498	0.532	0.504	0.476	0.06	0.05	<0.001
α_{Be}	0.678	0.690	0.670	0.688	>0.05	>0.05	>0.05
α_{Bi}	0.773	0.771	0.740	0.686	0.01	<0.001	<0.001

a total of 990 pairs of species. This number can be split into three groups: those pairs which differ in body size by a factor of less than 1.3 (first-order comparisons), pairs whose body size ratio is between 1.3 and $(1.3)^2 = 1.69$ (second-order comparisons), and pairs whose body sizes differ by a factor greater than 1.69 (third-order comparisons).

Overall niche overlap (summation alpha) decreases from first-order through third-order comparisons, and the trend is statistically significant. While two components of $\mathbf{a}$, α_H and α_{Be}, show no tendency to vary with the size difference between the species concerned, the two components α_V and α_{Bi} display significant trends. This is not surprising in the case of α_{Bi}, where a bill length comparison between the species is involved. The strength of this is sufficient to override lack of variation in α_{Be} and to produce a significant tendency in $\alpha_F = \alpha_{Be}\alpha_{Bi}$. In each case the tendency is the same: species of more similar body size show more similar foraging height distributions and employ more similar behaviors. To counteract this, there is a trend, not significant, for species of similar body size to be rather more different in habitat use than species of reduced body size similarity.

2. Linearly Distributed Resources. Bird communities may be supported by a few resources, each of large span, or by diverse resource types, each of short span. In the former case, the community's resources may be said to be linearly distributed, and species may evolve to form a serial array of replacements along the resource span. Clearly, the greater the tendency for resources to occur in long spans, the more species that will be linearly arranged along them, and the lower will be the mean overlap and the higher its variance in the community. In fact it is easily shown that, in a community of n species with a fixed and small degree of overlap between adjacent pairs along a single niche dimension ($\alpha = 0$ between nonadjacent pairs), the community mean overlap decreases

101

with species number $(= 2\alpha/n)$, and the variance in the overlaps increases at a decreasing rate $(= \alpha^2(1 - 2/n))$.

One resource axis on which species show good segregation is vertical foraging height. It might be supposed, therefore, that taller habitats could support more species along this dimension; mean (α_V) should decrease and var (α_V) increase with vegetation height. Figure 34 shows that this is only partly true. The expected relation holds for the habitats of shorter vegetation heights, up to around 10′ high. After this point both trends, a decreasing mean and an increasing variance,

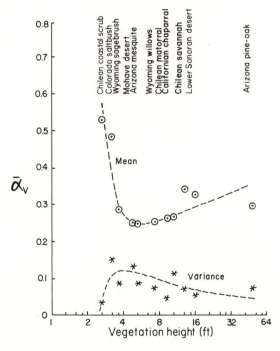

FIGURE 34. Mean community niche overlap in foraging height decreases with vegetation height up to around 6′, and then increases. Variance in foraging height overlaps does just the opposite.

appear to reverse themselves. The explanation for this may be that the taller habitats support a more diverse horizontal structural diversity at any one height, and thus species of different behaviors may co-occur at that height. Where two series of species that are vertical replacements or complements are superimposed on the vegetation height axis—for example one series of trunk searchers and one of foliage insect pursuers—the result would be as in Figure 34. At least two such series exist in Arizona pine-oak, whereas just one spans the Wyoming willows.

3. *Limiting Similarity, and the Number of Niche Dimensions.* The distributions of α for the eight study communities supporting more than five species are shown in Figure 35, while Figure 36 shows distributions of $\mathbf{a}$. These observed distributions of niche overlaps may be contrasted to distributions generated from random associations of numbers with the same range as niche overlap (0–1); these are called "expected distributions" for convenience and are also included in Figures 35 and 36. The expected distribution for summation alpha is generated by summing three numbers independently chosen from a flat distribution with range 0–1; by the central limit theorem, this produces a normal-looking curve, with its mean at 0.50. The expected distribution for product alpha is naturally quite different; the product of three numbers drawn with equal probabilities from the same flat distribution range 0–1 yields a curve with one-half of the products less than $(0.5)^3 = 0.125$, and half in the long tail above 0.125. The two figures show niche overlap distributions for both North and South American communities with between 12 and 20 species; three communities with just five species are omitted, as no significant patterns can be detected with such a small number (10) of niche overlap values.

Observed and expected distributions differ in several interesting ways. At the outset, it is somewhat reassuring to find that the differences are statistically significant (Chi-square

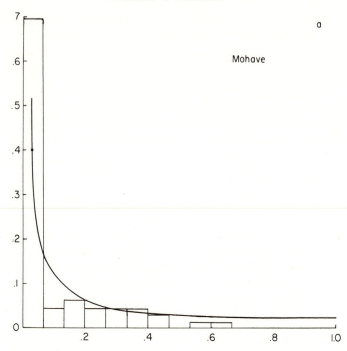

FIGURE 35. Distributions of niche overlaps (product alpha **α**) in six North American and two South American bird communities. Solid line generated from random products, histogram from actual data.

test, Figure 38) in all cases for α and almost all cases for **a**; we are ahead in that we at least have nonrandom number combinations. Looking first at **α** distributions, all communities deviate from expected in that they are short of high values. In fact there seems to be a tendency for truncation around 0.3, for scarcely any higher values are recorded (with the exception of the communities of low species number; see below). This gives considerable impetus to the notion of limiting similarity, and indicates its possible value around 0.5. The smaller communities (low species number, drawn first in the figures) also deviate from expected in their large excesses of low α values. As we look at increasingly larger communities, this excess

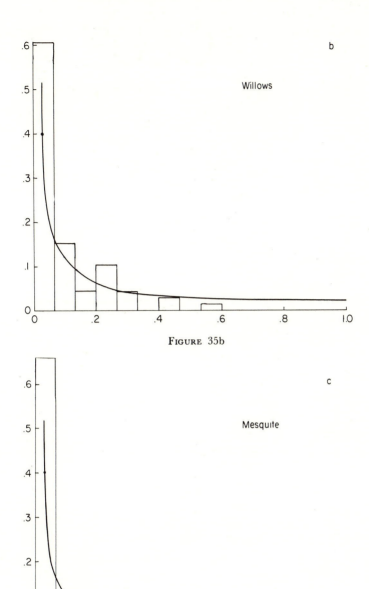

FIGURE 35b

FIGURE 35c

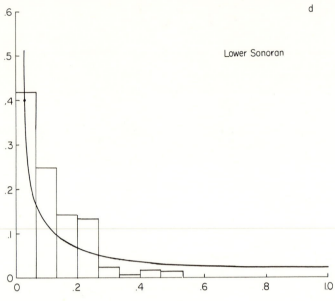

Lower Sonoran

FIGURE 35d

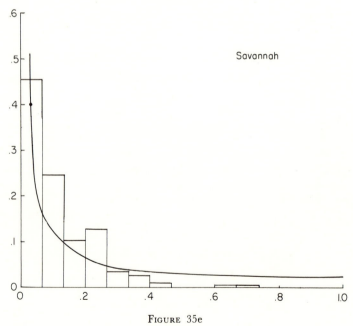

Savannah

FIGURE 35e

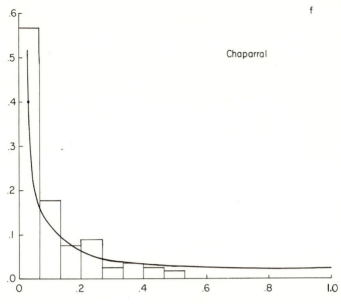

Chaparral

FIGURE 35f

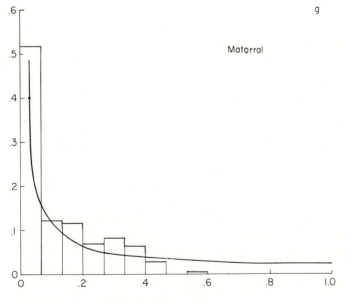

Matorral

FIGURE 35g

107

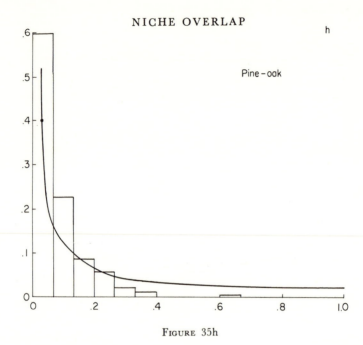

FIGURE 35h

becomes negligible at n species $= 16$, where excess frequencies occur only at intermediate α values, but becomes apparent again in larger communities ($n = 19$–20), in which excess frequencies at intermediate α are less obvious. These trends are clearly not fortuitous, for apart from being ordered with respect to community size, the California chaparral and its South American counterpart, Chilean matorral, show similar distributions, as do the Sonoran Desert community and a close Chilean equivalent, the savannah (open and tall, with saguaro-like and agave-like counterparts).

The **a** distributions are even more informative. Again there are trends which are most easily related to community size. The largest communities show a mode somewhat less than 0.5, and deficits at both high and low **a** frequencies. The smaller communities show large, obvious modes shifted back to **a** considerably less than 0.5. They display another feature that is quite remarkable, an apparent secondary hump at high

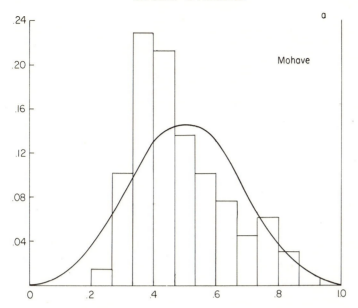

FIGURE 36. Distributions of niche overlaps (summation alpha **a**) in six North American and two South American bird communities. Solid line generated from random means, histogram from actual data.

overlap values 0.65–0.8. This might be suspect as artifactual, were it not for the fact that competition theory predicts the existence of bimodality in niche overlap distributions, and further predicts variation in the sizes and positions of the modes according to variation in resource utilization curves. MacArthur and Levins (1967) were the first to point out that species combinations that show segregation on some resource axis may be invasible by colonists that converge in resource use toward one of the other of the species which bracket it on the resource gradient. Roughgarden (MS) has extended this analysis, as illustrated by Figure 37. He considers two species on a resource gradient with particular resource utilization curves described by a kurtosis parameter t (the curves are platykurtic = flattened and thin tailed for $t > 2$, leptokurtic =

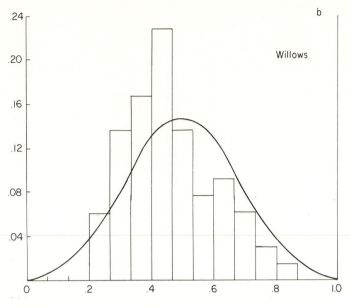

FIGURE 36b

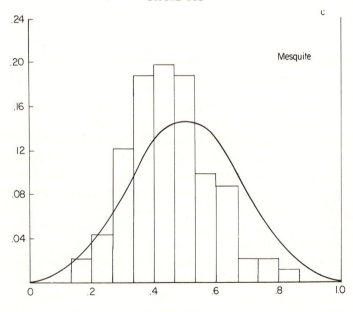

FIGURE 36c

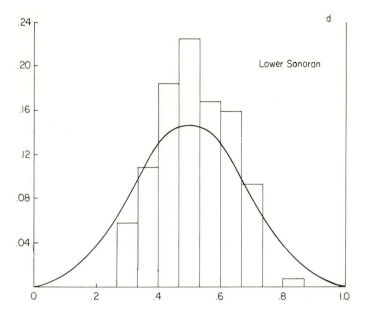

Lower Sonoran

FIGURE 36d

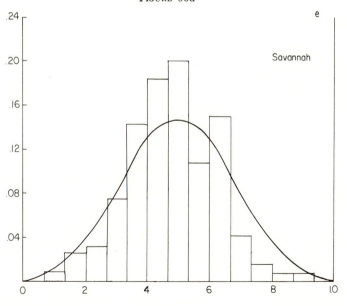

Savannah

FIGURE 36e

111

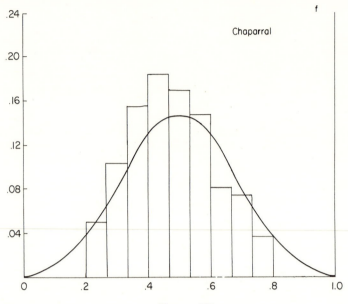

FIGURE 36f

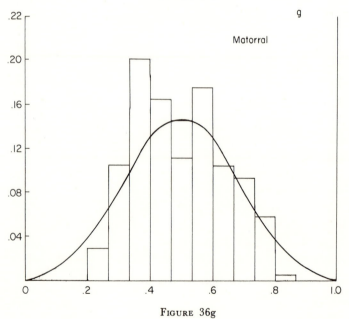

FIGURE 36g

112

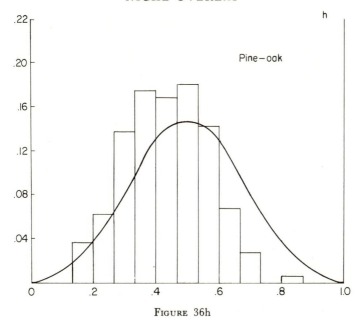

FIGURE 36h

pointed and thick-tailed for $t < 2$, and normal for $t = 2$). Suppose an invader arrives with an exactly intermediate resource utilization curve. Its success depends, for some value of t, on its niche overlap with the residents, which is a function of the separation distance of the resource utilization curves, and the ratio k/K of the invader's carrying capacity with the average carrying capacity of the residents. For $k = K$ and normal ($t = 2$) or mildly leptokurtic or Laplacian ($t = 1$) utilization curves, the colonist can invade only if its niche overlap with the residents does not exceed 0.55 (normal curves) or 0.65 (Laplacian curves). But if the invader has a 5% edge in carrying capacity, it can succeed with either moderate niche overlap (0.59 for normal curves and 0.75 for Laplacian curves) or with a high niche overlap (0.94 for normal curves and 0.92 for Laplacian curves).

113

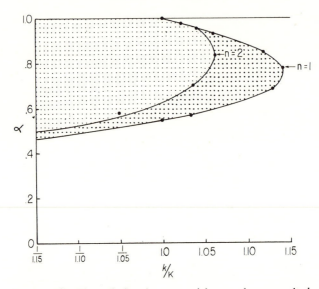

FIGURE 37. The relation between niche overlap α and the ratio of carrying capacities for invading species (k) and resident species (K). The guild of resident species is invasible if niche overlaps and carrying capacities fall outside of lightly stippled area for normal utilization curves, or outside of both lightly and heavily stippled areas for leptokurtic utilization curves. After Roughgarden, 1973.

R. May (MS) has further extended MacArthur's, Levins', and Roughgarden's analysis. He shows that the above results apply to an upper bound to α for a three-species coexistence at particular k/K ratios. There exists also a relevant lower bound, and coexistence is possible only for points in the k/K-α plane between these two bounds. Briefly, his results show that a) for $k/K = 1$, limiting similarity occurs at 0.544; b) for $k/K = \pm 1.05$, three-species coexistence is possible for all $\alpha < 0.59$ and for a small range $0.95 > \alpha > 0.94$; c) for larger k/K ratios, coexistence will occur only at increasingly small α values. In general, there is a large range of k/K over which coexistence occurs at small α, but an increasingly small range of k/K that maintains three-species coexistence at increasingly large α values.

Both 12-species habitats in Figure 36 appear to show a minor mode of high niche overlaps. These two habitats are

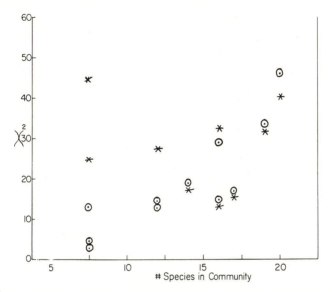

FIGURE 38. Value of Chi-square for the observed and expected niche overlap distributions in Figure 36, plotted against the number of species in the community.

distinguished from the others by being the highest in proportion of breeding migrants, in excess of 60%, and migrant species tend to be generalists with broad, flat utilization curves in comparison to year-round residents (see Chapters 4 and 6). The two sites are also lowest in vegetational complexity and resource span; perhaps an orderly ecological segregation is most likely to be realized in predictable and complex habitats in which a high proportion of the species are resident and competitively interactive year-round.

B. *Specialists and Generalists as Functions of Competitive Involvement*

Levins (1968) and later Vandermeer (1972) have discussed the covariance of the community matrix and its implications for the competitive structure of the community. Because of my wariness of scaling problems, I have assembled only symmetrical matrices, and so cannot consider their covariance. I can, however, use the row (= column) means of the matrices as estimates of the competitive involvement of the species concerned with the community as a whole (e.g. Table 3). It seems natural to label those species with low row/column means as specialists, as they exist in relative ecological

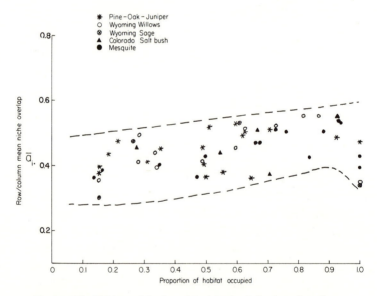

FIGURE 39. Habitat niche breadth (abscissa) is plotted against a measure of competitive involvement of each species with the remainder of the community (large for generalists, small for specialists). (a) shows a broad trend for generalists to be commoner in the habitat than are the specialists in five climatically unpredictable habitats, but this trend does not occur in (b) three climatically predictable habitats.

116

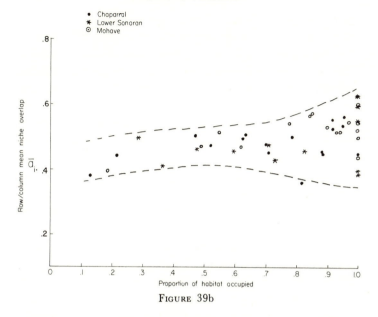

FIGURE 39b

isolation, whereas others with high row/column means reflecting high competitive involvement can be described as generalists. We can use these measures to investigate the sorts of species that are common in the community, the "dominants." There are two alternatives; species are common because they are generalists and use a wide variety of the habitats or resources (MacNaughton and Wolf, 1970), or are common because they are specialists, and their speciality is common in the habitats. As a measure of success, I shall use the proportion of the study area occupied by a species, because an alternate measure, the number of pairs/unit area, is subject to variation with body size differences.

The results are shown in Figure 39. The eight North American study areas are divided into two groups (of 56 and 45 species total) according to overall resource predictability (R_V, T_V, FFD). Figure 39a shows that the widely distributed species in the unpredictable habitats are generalists; indeed, generalism seems to assure success. While there are only a few rare generalists, there are plenty of rare specialists, and of these only the aerial feeders attain more than 50% habitat coverage. I must add, though, that many of these specialists are habitat specialists; this association between the variables notwithstanding, we might expect more rare generalists than we observe, yet they are all common. Notice that these are the species farthest below carrying capacity due to their heavy competitive associations; nevertheless, they are the most common species in the habitat.

The more predictable habitats show a slightly different pattern (Figure 39b). Here specialists are as likely to be common as rare, and while most of the dominants remain generalists, the specialist representation is greatly increased.

Two widespread and somewhat specialized types, hummingbirds and thrashers, illustrate these effects quite well. Five different thrasher species occur in five different study areas and a thrush counterpart occurs in a sixth area, the willows. These species tend to be quite common in the three more predictable habitats, (average $B_H = 0.863$). In the three more uncertain habitats, however, they are much rarer (average $B_H = 0.311$), even though there is only a slight reduction in mean row niche overlaps, from 0.47 to 0.42. The more specialized hummingbirds are found in all six North American sites with vegetation over $3'$ high. In the three most predictable of these, they are quite common (average $B_H = 0.695$) but not too specialized ($\alpha_{i\cdot} = 0.47$). In the more uncertain habitats, where the species stand further apart from their competitors ($\alpha_{i\cdot} = 0.39$), they are rarer (average $B_H = 0.431$).

III. BETWEEN-COMMUNITY VARIATION IN NICHE OVERLAP

A. Average Niche Overlaps in Communities

1. Consistency in Mean Overlap Values. Community mean values in niche overlaps and their components are given in Table 5

TABLE 5. Mean community niche overlap values.

Study area	Mean niche overlap		Niche overlap components			α_F components	
	a	α	α_H	α_V	α_F	α_{Be}	α_{Bi}
Wyoming willows	0.458	0.093	0.524	0.254	0.598	0.655	0.835
Wyoming sagebrush	0.431	0.090	0.473	0.286	0.534	0.661	0.793
Colorado saltbush	0.471	0.085	0.421	0.484	0.509	0.768	0.664
Mohave desert	0.484	0.092	0.751	0.254	0.446	0.616	0.728
California chaparral	0.479	0.092	0.717	0.268	0.478	0.687	0.695
Sonoran desert	0.517	0.109	0.772	0.328	0.418	0.640	0.664
Arizona mesquite	0.448	0.075	0.622	0.249	0.462	0.630	0.738
Arizona pine-oak	0.434	0.075	0.480	0.299	0.522	0.701	0.738
Coastal scrub, Chile	0.448	0.105	0.384	0.531	0.423	0.663	0.686
Matorral, Chile	0.509	0.115	0.750	0.261	0.513	0.680	0.751
Savannah, Chile	0.480	0.112	0.512	0.341	0.586	0.762	0.766

for 11 study areas. Of interest here is the consistency of the average niche overlap, in particular of summation alpha a, among the different communities. The values vary only 16% in spite of the fact that the habitats studied vary drastically in vegetation height and in climate, support from 5 to 20 bird species, and span two continents. Variation in product alpha α, while less in absolute magnitude, is proportionally much greater.

Overall niche overlap has three components, α_H, α_V, and α_F. Of these, habitat overlap α_H shows by far the greatest variation among communities (range 0.388), vertical overlap shows a reduced variation (range 0.282), and overlap in food/feeding behavior α_F remain relatively constant (range 0.180).

The variability in niche overlap components appears to reflect the ecological plasticity of the birds in the communities, and parallels the niche shifts observed after more drastic changes in competitive environment from mainlands to islands. Several studies have shown that bird species readily and commonly broaden their habitat use on islands, less commonly change their vertical foraging distributions, but seldom change their foraging behavior in any obvious way, suggesting its innate control and strong dependence on morphology. The other correlate of diet, bill structure, appears to be the more adaptable of the two components of α_F.

Competition theory has predicted upper bounds for α in coexisting species that vary with their relative carrying capacities. My figures (Table 5) are all within the limits to α expected when the majority of the species concerned are of roughly equal carrying capacity. Earlier I had found (Cody, 1968) that grassland bird species also had achieved a fairly uniform degree of ecological segregation (average separation 135.0/300, average community niche overlap 0.550). My figures (overall $\mathbf{a} = 0.469$, $\alpha = 0.095$) fall between those of Pianka (1969), who finds $\mathbf{a} = 0.42$ and $\alpha = 0.07$ in an Australian lizard community, and Pico et al. (1965), cited by Pianka, who measure $\mathbf{a} = 0.65$ and $\alpha = 0.26$ in Puerto Rican Drosophila species, and are close to those of Orians and Horn (1969), who found niche overlaps among blackbird species in Washington to average 0.51.

2. *Interrelations Between Niche Overlap Components.* It is useful to see how variation in overall niche overlap is attributable to variation in its various components. Figure 40 plots $\mathbf{a}$ as a function of α_H. Increasing habitat overlap effects an increase in overall niche overlap despite the fact that the average values of the other two components of $\mathbf{a}$, $(\alpha_V + \alpha_F)/2$, decrease with increasing α_H. These results show that communities represented by points at the upper right of the figure support species that are similar in their use of habitat, but feed at different heights and on different sorts of foods, and are distinguished from other communities in which species are much more similar in vertical foraging distributions and in foraging behaviors, but where habitat overlap is sufficiently low that overall niche overlap is reduced. We next investigate

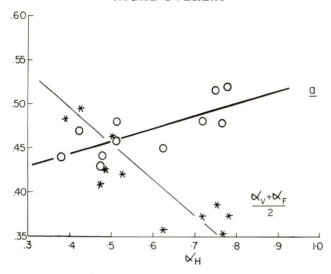

FIGURE 40. The relationship between summation alpha a (open circles) and the average of α_V and α_F (asterisks), and habitat niche overlap α_H. Increases in overall niche overlap are entirely due to increasing overlap in habitat use.

environmental differences that might explain these community differences.

B. Climatic Factors and Niche Overlap

1. Stability and Predictability of Resources. The idea that the degree of tolerable niche overlap should be related to resource predictability and/or stability has permeated the ecological literature for some time. It was recently made explicit by Slobodkin and Sanders (1969). May and MacArthur (1972) investigated the question theoretically and reached the conclusion that species reach maximal niche overlaps that are relatively insensitive to environmental fluctuations (i.e. a thousandfold change in environmental variance produces only a 50% change in limiting selectivity) provided that these fluctuations are not extreme, or arctic.

The relation between community and climatic predictability

and seasonality was tested using step-wise regression analysis and the climatic variables already mentioned (Table 1). One climatic variable, rainfall variance R_V, turned out to be consistently the most powerful predictor of mean niche overlaps **a** and α and their components, and was most often selected as the variable which "explained" most of the variance in the dependent. variables. The second most important climate variable turned out to be the number of frost-free days per year, an index of seasonality and of the permanence of the community. Lastly P, the probability of unusual weather, and T_V, mean temperature predictability, show some capacity for variance reduction, but the two seasonality measures $R(M - M)$ and $T(M - M)$ are both poor indicators of niche overlaps. Figure 41 shows how these climatic variables are related to **a**, α, and α_H in stepwise regression, and plots **a** sequentially against functions $f(R_V)$, $f(R_V, T_V)$, and $f(R_V, T_V, P)$ which minimize its variance. This last function, Figure 41f, orders the communities with respect to predictability of breeding season rainfall, and communities that enjoy high predictability of resources are those which show the highest niche overlaps. As indicated above, it is increase in habitat overlap that produces the higher overall niche overlaps in predictable environments, and overlaps in α_V and α_F change in the opposite direction. This trend presumably terminates in the most predictable communities with α_H at its upper limit of 1.0, approached asymptotically, whereafter further increase in **a** must be accomplished by increases in α_V and α_F. These results lead to the conclusion that environmental uncertainty associated chiefly with unpredictable rainfall is more likely to affect a segment of the food resources than a subsection of the habitat. To combat this uncertainty, species have evolved to a more monopolistic habitat use in unpredictable climates, and in these habitats they behave as generalists, in similar ways to other species using other subsections of adjacent habitat in an equally exclusive way.

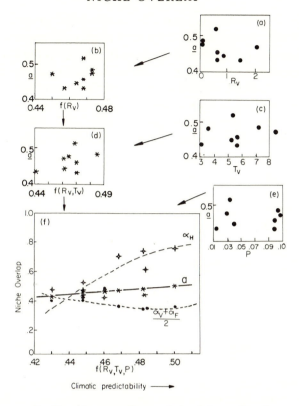

FIGURE 41. Step-wise regression analysis of three climatic predictability variables with niche overlap. Niche overlap increases with climatic predictability (80% variance explained), due to increase in overlap of habitat use α_H.

We can now comment further on the question of which measure of niche overlap is more appropriate, summation alpha or product alpha. Three climatic predictability measures account for 63% of the variance in **a**, but only 43.8% of the variance in **α**. Predictability estimators are generally much more successful than seasonality estimators in accounting for niche overlap variance, and again for seasonality measures **a** behaves in a more orderly fashion than does **α** (35% variance explained, versus 22% in α). The habitat overlap component of niche overlap proves to be most sensitive to resource predictability (90% variance explained), and equally responsive to seasonality factors (89% variance explained). The other two components respond to the extent that between 34% and 74% of their variance is accounted for by changes in either resource predictability or resource seasonality. These results are summarized in Figure 42; they support the conclusions that climatic predictability permits larger niche overlaps

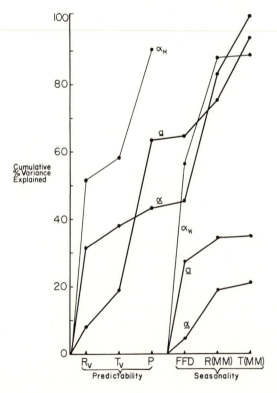

FIGURE 42. The power of six climate variables in accounting for variance in two niche overlap measures (heavy lines) and niche overlap components (light lines).

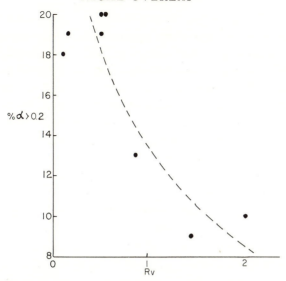

FIGURE 43. High values of niche overlap (product form) occur only where rainfall variance is low.

and that the different niche dimensions are considerably interdependent in these bird communities.

Although product alpha is quite poorly correlated with resource predictability in its average value, it is clear that its distribution has been shaped by climate in each study area. Figure 43 shows that the proportion of α values in excess of 0.2 is largest in the most predictable habitats, and is low in unpredictable areas. This in turn may indicate that selection is acting upon a threshold value for product alpha rather than on its mean value.

C. Community Stability

Recently ecologists have wondered if some simple property of the community matrix can be used as an index of the stability of the community whose interactions it summarizes. By stability is meant the resistance of the community to changes in species number or relative abundance, or its rapid return to the previous equilibrium population sizes following perturbations. Levins (1968) pointed out that the determinant of A must be positive for the persistence of all species, but the value of the determinant seems not to be useful as a stability index. In the eight North American study areas, the determinant values, while all positive, were very strongly correlated with the size of the community, being largest in the largest communities $(r = 0.9)$, an effect which appears to swamp any predicted relation between community stability and resource stability or predictability.

More recently it was suggested that the eigenvalues of the community matrix will reflect the stability of the system (May and MacArthur, 1972), and that in particular the value of the smallest eigenvalue, λ_{min} will be proportional to the "environmental noise level," or to environmental uncertainty. It will eventually be possible to test this model with the bird community data in Appendix A.

IV. SPECIES DIVERSITY AND SPECIES PACKING

For resource-limited species, the total niche or resource space is filled by the n resident species showing some average niche size and some average characteristic niche overlap. Thus the species diversity D_s in the community was written by MacArthur (1972) as

$$D_s = D_r(1 + c\alpha)/D_u,$$

or, in my notation, number of species

$$n \propto \prod_{k=1}^{r} R_k\alpha \Big/ \prod_{k=1}^{r} B_k$$

where ΠR_k measures niche space, α mean niche overlap, and ΠB_k niche size, the product of niche breadths in r dimensions. MacArthur and associates have shown that habitat structure, in any particular geographic locale, approximates resource span or niche space, ΠR. Above we showed that resource predictability, as estimated by climate predictability, allows reduced niche overlap; in addition, it may affect niche breadths, and therefore niche sizes, but in somewhat ambivalent ways. A third factor is the productivity of the habitat, which, according to both theory (MacArthur, 1971) and ecological intuition, should be positively correlated with the closeness of species packing. Thus, we have the option of relating species diversity directly to the environmental measures of resource span R, resource productivity and resource predictability, or less directly to resource span, niche overlap and niche breadth. Table 6 gives species diversity measures, niche measures, and three resource measures. Resource span is measured by the equitability and extent of the vegetation distribution in the vertical plane; resource productivity is estimated from actual evapotranspiration (Rosenzweig, 1968b), which in turn is estimated from climatological data; and resource predictability is measured as before by breeding season rainfall variance R_V. It is well known that bird species diversity and habitat structure are related, and using the Table 6 data, 81% of the variability in the dependent variable BSD is accounted by habitat structure. In addition and in just the way the theory suggested, bird species diversity is positively correlated with resource productivity AE $(r = 0.32)$ and negatively correlated with resource unpredictability $R_V(r = -0.26)$, although neither relation is by itself statistically significant. Incorporation of the second two resource variables into a regression equation accounts for a further 9% of variability in bird species diversity: $BSD = 1.33FHD + 0.00143AE - 0.104R_V$, $(F = 11.16$, $p = 0.02)$. Because in my data the number of species at a site is strongly correlated with the bird species diversity there $(r = 0.96)$, similar treatment of number of species n as a dependent variable is just as successful (91% variability explained), but interestingly enough from the theo-

Table 6. Estimates of resource span, resource productivity and resource predictability in eight North American study areas.

Study area	# species	BSD^1	Resource span[2] (FHD)	Resource productivity[3] (AE)	Resource unpredictability[4] (R_V)
Wyoming willows	12	2.26	0.689	290	0.504
Wyoming sagebrush	5	1.34	0.056	290	0.504
Colorado saltbush	5	1.35	0.081	275	2.036
Mohave Desert	12	2.03	0.687	177	0.154
California Chaparral	17	2.42	0.435	473	0.099
Sonoran desert	16	2.66	0.830	291	0.541
Mesquite scrub	14	2.10	0.677	439	0.868
Pine-oak woodland	20	2.71	1.018	352	1.465

[1] Measured by the information-theoretic formula *fide* MacArthur and MacArthur, 1961.
[2] Measured by "foliage height diversity" = habitat structure, *fide* MacArthur and MacArthur, 1961.
[3] Measured by actual evapotranspiration, which is estimated by Turc's (1955) formula $AE = P/[0.9 + (P/L)^2]^{1/2}$, where P = annual precipitation in mm, $L = 300 + 25T + 0.05T^3$ and T = annual mean temperature.
[4] Estimated by breeding season rainfall variance R_V.

retical standpoint, n is less dependent on FHD (75% vs. 81%) and much more sensitive to resource productivity (16% vs. 7%). As MacArthur (1971) suggested, more species can be packed into more productive habitats, as the unused resource production, when combined species utilization curves are subtracted from the production curve, amounts to enough to support individuals of an additional species.

CHAPTER FOUR

The Variable Competitive Environment

The great majority of bird species occurs in a variable competitive environment. This is because species encounter different sets of competitors in different habitats, different parts of the geographic range, and at different times of the year. Thus, it is not surprising that ecologists have looked for, and have generally found, niche shifts that parallel shifts in the competitive environment.

The effects of changing competitive regime can be predicted if the species utilization curves are known. Given a constant resource gradient, a species from the middle of a serial array (Figure 19) would be expected to shift its utilization curve to the right if competitors are removed from the right, to the left if left-side species are absent, and to show an overall increase in niche breadth in a generally impoverished environment. Change in niche size or position following a reduction in competitor diversity is termed *competitive release;* the term *character displacement* has been used to describe a divergence in morphological or ecological characteristics in sympatric species from their more similar character states in allopatry. Below, we discuss under what sorts of conditions the classical character displacement might evolve, and then discuss two pairs of contrasting environments, mainland/island and summer/winter, in which wholesale changes in competitive environment occur.

VARIABLE COMPETITIVE ENVIRONMENT

I. CHARACTER DISPLACEMENT

A. The Nature of Character Displacement

A "classical" character displacement hypothetically involves a pair of potentially competing, often closely related, species that are only marginally sympatric. In their zones of allopatry, they are similar in some morphological or ecological character, but, when and where they co-occur, this character is displaced in each, in opposite directions, away from its value in allopatry. Grant (1972) has recently reviewed the evidence for this form of character displacement in several examples commonly cited in the literature, and finds it weak. He points out that very often the displaced character values in sympatry a) vary clinally over both the sympatric and allopatric parts of species ranges, or b) differ from character values in allopatry in just one of the two species involved, or c) represent a more primitive original condition, and that the derived condition is convergence in character values between the allopatric populations. These variations might be considered, by definition, to fall outside of the term character displacement. However, we are interested primarily in understanding alternative outcomes of competition. One broad category of results is greater ecological difference (because of either altered morphology or altered behavior) between the sympatric populations of two species than between their allopatric populations, and character displacement is an acceptable term for the phenomenon.

We can note that the extent to which a broadly defined character displacement differs from the restricted classical definition parallels the extent to which sympatric populations fail to maintain genetic isolation from their conspecific allopatric populations. Thus variation between island and mainland populations is more likely to be discontinuous or step-clinal, whereas variation between continuous mainland populations will likely be simply clinal. Also, island populations often face simpler competitive environments and are far more often derived from mainland populations than vice versa. Thus the derived population is likely to display values intermediate between mainland (displaced) values.

It is also likely that the extent to which clines in character variation are stepped rather than smooth is related to discontinuity in the competitive environment. Wholesale changes in numbers or types of competitors could abruptly change optimal character values, and these new values can be attained if genetic variation and isolation both exist. These large-scale changes in competition are most obviously expected between island and mainland populations, but also between summer and winter environments. Discontinuous character variation is further expected in simple communities, or guilds, where the bulk of the restraint on realized niches comes from just one or two competitor species. In such cases of "simple competition," the addition or subtraction of just one of these competitors can drastically alter optimal niche position (or size). But where a species has numerous competitors, as perhaps generalists might, a one-or-two-species change in competitor number is unlikely to produce much change in optimal character value. This is called "diffuse competition" (MacArthur, 1972), and will probably mask adaptive niche shifts in complex communities of generalized species.

B. Niche Shifts on Resource Gradients

Field evidence now exists to document abundantly changes in both niche position and niche size with altered competitive regime. As I wish to spend time below on two specific categories of large-scale turnover in competitors, this section will be brief. One of the valuable functions of this evidence is to show that often no compelling innate-genetic or physiological constraint restricts a species to a particular section of the resource gradient, but rather that its position is flexible, and is determined by the restraints of its competitors.

Several types of niche shift may come about rapidly, through altered behavior rather than by genetic change. Thus a species that was confined because of competitors above it to the lower slopes of a mountain range can expand its altitudinal range over the whole of an isolated peak which it occupies alone, and probably does so with little if any physiological change. Diamond (1970a) gives several fine examples of character displacement in altitudinal range on islands in the New Guinea region. In the same way species that segregate by habitat may rapidly occupy additional habitat types left vacant by absent competitors. This change, to a broader habitat use in competitor-poor environments, is perhaps the most prominent difference between island birds and their mainland counterparts, and has been widely noticed (Lack, 1971). In contrast to the usual pattern, the Canary Islands have two chaffinches (*Fringilla* spp.) as opposed to one on the European mainland. On the island, the two segregate by the type of woodland they live in, but on the mainland the single species breeds and is abundant in the coniferous forest from which it was excluded on the Canaries.

Vertical foraging ranges can also shift quickly, and often expand in response to reduced competitor pressure. Crowell (1962) reported that both the distribution of foraging heights and the use of foraging sites (foliage, ground, bark) had changed considerably in the most common species, cardinal, catbird, and white-eyed vireo, on the island of Bermuda, but in only one species were they significantly larger. The flycatchers in Figure 28 demonstrate the same flexibility. More examples will be given below.

Changes in diet may require a change in bill structure, in body size, in feeding behavior, or in all three; certainly in the case of the first two and apparently also in the last, an alteration of genotype is required. Diet may be expanded by increased bill or body size, and such changes are common in differentiated island populations. But species that feed from

132

the larger end of a prey size series, such as higher predators, may decrease in body size when they colonize habitats from which their smaller competitors are absent. The California Channel Island foxes (*Urocyon littoralis*) are tiny compared to their mainland counterparts (*U. californicus*), presumably because raccoons, some skunks, and kit foxes are absent. Most herbivores, omnivores, and insectivores show larger body sizes on islands (voles on small British islands such as Orkney and Skomer—Southern and Linn, 1964; bears on the Queen Charlotte and Kodiak Islands—Burt and Grossenheimer, 1952). Lizards are larger on most islands (e.g. *Klauberina* on the California Channel Islands is two to three times the weight of its mainland relative *Xantusia*), but a reduced productivity may favor reduced body size despite reduced competitor diversity (Schoener, 1970). Also the California Channel Islands used to support a pygmy variety of elephants. Exact predictions of niche shifts are possible only when the resource states and the utilization curves of the species that depend on them are known.

C. Cross-overs in Displacement Patterns

Occasionally species adjacent in serial array on a resource gradient occur elsewhere with their relative positions reversed. This illustrates forcibly that more importance must be attached to species' *relative* positions on the resource gradient rather than to the *absolute* position each occupies (Figure 44). Two examples come from seabirds. Of the two common breeding cormorant species of the Olympic Peninsula, Washington state, the smaller 27″ pelagic cormorant *Phalacrocorax pelagicus* is darker with a more glossy plumage, feeds further offshore, and breeds on steeper cliff faces than the second, larger species, the 36″ double-crested cormorant *Phalacrocorax auritus*. The pelagic cormorant breeds later than the double-crested, such that the young of the two species scarcely overlap in time on their breeding islands (Cody, 1973a). In

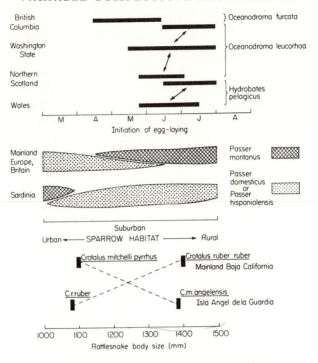

FIGURE 44. Crossovers in character displacement occur in (a) the breeding seasons of small oceanic petrels, (b) the habitats of sparrows *Passer,* and (c) in rattlesnake body size.

England also there are two breeding cormorants, close ecological counterparts although different species; again the smaller one, the 30″ green cormorant *Phalacrocorax aristotelis,* is darker and glossier, feeds further offshore and nests on more inaccessible cliffs than its larger congener, the 36″ great cormorant, *Phalacrocorax carbo.* But in England it is the smaller green cormorant which breeds early, and the great cormorant begins to breed four to five weeks later. The second example involves the small petrels discussed in Chapter 1; two species are largely separate in their breeding seasons where they occur together off the British Columbia coast, but Leach's

134

petrel occurs alone on the Olympic Peninsula coast and begins breeding at intermediate season. Leach's petrel is the late breeder in British Columbia, and this species also occurs on islands in northern Scotland together with another, but different, small petrel species, the storm petrel *Hydrobates pelagicus*. Here the two petrels also show a segregation by breeding season, but now Leach's petrel is the early breeder in the sequence rather than the late one (Figure 44).

An example of a crossover in habitat use is provided by the European sparrows (*Passer*) studied by Hartmut Walter (pers. comm.). Typically in Europe (and elsewhere, see Chapter 5), two species of *Passer* segregate along an urban-suburban-rural habitat gradient; *P. montanus* occupies the greener, rural end of the gradient, and *P. domesticus* or one of its semispecies *P.d. italiae* or *P. hispaniolensis* lives in the city centers and suburbs. On the island of Sardinia, however, *P. montanus* is restricted to the downtown and port area of the major city of Cagliari; the more typically urban sparrow *P. hispaniolensis* is displaced to the suburbs and countryside. Undoubtedly the order of colonization of Sardinia by these sparrows is responsible for the habitat switch.

A final example comes from the South American bird family Rhinocryptidae. There are four genera in Chile—*Pteroptochos, Scelorchilus, Eugralla,* and *Scytalopus*. There are two species in each of the first two genera, one in each of the second two; the congeneric species are almost entirely allopatric. In some parts of the country, the species are convergently similar in body size, but, where they segregate by size, the order is according to that written above, with *Pteroptochos* the largest. All four genera (four species) occur on the island of Mocha, where they segregate by size. And only on Mocha is the representative of *Scelorchilus* larger than *Pteroptochos,* and by just the same ratio as the larger *Pteroptochos* exceeds *Scelorchilus* on the mainland!

Such crossovers occur in other animal groups, and a good

example comes from the rattlesnakes *Crotalus*. The red diamond rattlesnake *C. ruber* and the speckled rattlesnake *C. mitchelli* co-occur extensively in southwestern North America. The two differ in average size, with *C. ruber* 27% longer (twice as heavy) on average. *C. mitchelli* apparently reached the Gulf of California island of Angel de la Guardia long ago, for it has differentiated there into a much larger form, *C.m. angelensis* (contrary to the common pattern of size reduction with competitive release in these snakes; Klauber, 1963). *C.r. ruber* has also reached Angel, but presumably much more recently than *angelensis,* for it is taxonomically undifferentiated from the mainland form. This island population of *ruber* does differ from mainland forms, however, in body size, for it is considerably smaller (Schmidt, 1929). Thus the positions of the two species in the size sequence have been reversed (Figure 44).

II. ISLAND COMMUNITIES

A. The California Channel Islands

One of the most detailed studies contrasting mainland to island communities is Yeaton's (1972) study of chaparral birds in the Santa Monica Mountains of Los Angeles County and on Santa Cruz Island, 18 miles off the California coast. On study areas around 7 acres in size, he found 17 species on the mainland and 12 species on the island, with 9 species in common. Thus the island community is not terribly impoverished and nor will it be completely restructured, so that niche shifts and species replacements will be easy to measure. Habitat structure was very similar between the two locations, bird predators occurred in both places, and both the size diversity and the species diversity of the bird's insect prey were the same, although insect biomass was 20% greater on the island, perhaps due to its more equable climate.

Several of the most conspicuous birds of mainland chaparral

are absent from Santa Cruz Island (see Table 7). Two of these, Anna's hummingbird and Nuttall's woodpecker, can be matched with reasonable precision to ecological counterparts on the island, Allen's hummingbird *Selasphorus sasin* and

TABLE 7. Density changes and niche shifts between mainland and island California chaparral communities (from Yeaton, 1972).

Guild	Mainland density (pr/ac)	Main feeding zone	Niche breadth Bv	Island density (pr/ac)	Main feeding zone	Niche breadth Bv
Foliage insectivores						
Orange-crowned warbler	0.07	>15'	1.59	0.55	8'-15'	2.62
Common bushtit	0.20	6"-8'	2.23	0.24	6"-8'	2.64
Hutton's vireo	0.14	>15'	2.50	0.55	6"-8'	2.93
Bewick's wren	0.61	6"-8'	2.70	1.27	6"-8'	1.76
Wrentit	1.77	6"-8'	1.23	absent		
Plain titmouse	0.20	6"-8'	2.93	absent		
Blue-gray gnatcatcher	absent			0.63	8'-15'	2.39
	2.99		2.20	3.24		2.47
Nectivores						
Anna's hummingbird	0.48	8'-15'	1.99	absent		
Allen's hummingbird	absent			0.63	8'-15'	1.99
Sallying flycatchers						
Ash-throated flycatcher	0.20	aerial	1.00	0.16	aerial	1.00
Ground feeders						
Black-headed grosbeak	0.10	8'-15'	2.20	0.10	8'-15'	2.20
Rufous-sided towhee	0.54	ground	2.20	absent		
Scrub jay	0.27	6"-8'	3.77	1.11	6"-8'	2.65
Brown towhee	0.20	ground	1.49	absent		
California thrasher	0.48	ground	2.72	absent		
California quail	0.12	ground	1.00	absent		
	1.71		2.23	1.21		2.43
Woodpeckers						
Red-shafted flicker	0.07	ground	1.55	0.07	ground	1.55
Nuttall's woodpecker	0.07	15'	1.00	absent		
Acorn woodpecker	absent			0.08	15'	1.62
	0.14		1.28	0.15		1.59
Aerial feeders						
White-throated swift	++	aerial	1.00	++	aerial	1.00
Total densities and mean niche breadth:	5.52		1.95	5.39		2.03

acorn woodpecker. An additional three prominent mainland species—brown towhee, California thrasher, and wrentit—have no obvious island replacements. Two additional species, rufous-sided towhee and California quail, were present on the mainland site, but, although they occur on the island, they were scarce or absent from the study area. Two species that are marginal in mainland chaparral and are more typical of oak woodland—orange-crowned warbler and Hutton's vireo—become abundant in island chaparral, and three more species restricted to oak woodland on the mainland—acorn woodpecker, Allen's hummingbird, and blue-gray gnat-catcher—are common in the island plot. The two most dramatic density changes are observed in Bewick wren, which doubles in density in apparent response to the absence of wrentit (the most common bird of mainland chaparral) and of plain titmouse, and in the scrub jay, which is a remarkable four times denser on the island. This increase is in obvious response to the absence from the island of four common mainland ground-feeding birds; the island jays spend much more of their time feeding on the ground, and have even developed longer, almost thrasher-like bills with which they dig in leaf litter. The net result of these shifts and replacements is an increased density in the species in common between island and mainland, from 1.66 to 4.05 pr/ac, to compensate for reduced numbers of species restricted to the island versus restricted to the mainland (eight mainland-restricted species total 3.86 pr/ac, whereas three island-restricted species total 1.34 pr/ac); overall bird densities approach equality at 5.52 pr/ac (mainland) versus 5.39 pr/ac (island).

The striking increases in densities of jays, wrens, warblers, and vireos in the island community are presumably a result of the increased resource density available to these species in the absence of their mainland competitors. In the foliage insectivores, both orange-crowned warbler and Hutton's vireo shift their vertical foraging zones down from 15' to below 8', presumably to use niche space vacated by wrentits. Their niche breadths B_v are rather broader on the island, indicating that both still do some feeding in higher foliage levels. Bewick's wren already feeds in the wrentit zone; thus its vertical foraging range remains the same, and its niche breadth actually decreases in presumed response to much increased food abundance within its foraging height range. The blue-gray gnatcatcher joins orange-crowned warblers in the higher foliage levels now abandoned by Hutton's vireo.

The situation in the ground-feeding omnivores is even more striking. All of the food resources shared among five mainland species now go to the scrub jay, which increases in abundance correspondingly. This high food abundance at and near ground level has permitted a reduction in niche breadth in the jay, which now most profitably concentrates its feeding activity at these lower levels. Note that, although density compensation is pronounced, overall bird density on these resources on the island is only 70% of that on the mainland.

These new island equilibrium densities can be predicted via competition theory. The mainland community yields a community matrix **A** and equilibrium densities X_i, from which carrying capacities K_i can be calculated from the matrix equation $\mathbf{K} = \mathbf{AX}$. If we assume that carrying capacities and competition coefficients will be unchanged in the similar island habitat, new island equilibrium densities Y_i can be calculated, with provision for the new island species, from the matrix equation $\mathbf{A^{-1}K} = \mathbf{Y}$. Table 8 shows just how close is the match between predicted and observed equilibrium densities on the island. Yeaton concluded that, through niche shifts and niche expansion in both foraging height and habitat use, the fewer island species use close to the same total resources as the larger number of mainland species. In fact, when absent mainland species are now "introduced" into the niche-shifted and niche-expanded island community, via the matrix equations, negative equilibrium values result, indicating that, even if these

TABLE 8. Comparison of predicted versus observed island densities on the basis of mainland competition coefficients and carrying capacities.

Species	Mainland density	Island density Predicted	Observed
Scrub jay	0.27	1.00	1.11
Orange-crowned warbler	0.07	0.50	0.55
Hutton's vireo	0.14	0.60	0.55
Bewick's wren	0.61	1.28	1.27

missing species were now to reach Santa Cruz Island, they would fail to become established there!

Diamond (1970b) and MacArthur *et al.* (1972) have suggested that the extent to which density compensation can occur in island faunas depends on the appropriateness of the colonizing species to the island habitats. All of the Santa Cruz chaparral birds are birds of mainland chaparral or of adjacent chaparral-oak woodland mixtures, and thus density compensation is nearly complete. Yeaton found, however, that the birds living in the relict pine forest on Santa Cruz Island are mostly species typical of chaparral (the three most common species there are Bewick's wren, scrub jay, and orange-crowned warbler, none of which occurs in the mainland pine forest). So despite the fact that the reduction in species number from mainland to island pines (14 spp. vs. 10 spp. = —29%) is the same as from mainland to island chaparral (17 spp. vs. 12 spp. = —29%), density compensation is much less complete in the island pine community (4.94 pr/ac vs. 6.71 pr/ac, = 74%, compared to 98% in chaparral).

B. Wrens on North Atlantic Islands

For reasons that are not at all clear, wrens (Fam. Troglodytidae) turn out to be remarkable colonists. Not just some wren species, for virtually all of the north temperate forms have reached offshore islands. In the northeast Pacific, winter wrens (*Troglodytes troglodytes*) occur on islands from the Pribilofs off Alaska to tiny Solander Island offshore from Vancouver Island. Almost all the islands off southern California and northern Baja California have wrens, either one or two of the four species on the adjacent (mainland) coast (house wren, rock wren, canyon wren, and Bewick wren) or their derivatives. So do islands in the Gulf of California, the Panamanian Pearl Islands (a wren is the commonest bird on Puercos), and even the tiny islets of Gatun Lake, Panama Canal Zone, support abundant wren populations and in some cases almost nothing else. In the Caribbean, only the Lesser Antilles support wrens, but the Northern Atlantic record is particularly striking. *T. troglodytes* occurs on the Shetland and Orkney Islands, Fair Isle, the Inner and Outer Hebrides, St. Kilda, the Faeroe Islands, and Iceland. Six subspecies have been named from these islands, differing from the British mainland *T. troglodytes* chiefly in their larger size and coloration.

English wrens are woodland birds. In Yorkshire they live in the understory of insect-rich beech forests with around a dozen other insectivorous passerine bird species; each pair has an average brood size of 6.1 young (Cody and Cody, 1972). Wrens of the mainland subspecies on the Isle of Skye, Inner Hebrides, live in low and open hazelnut and birch woodlands, which they share with six other passerines, and wrens on Iceland live in a similar dwarf woodland habitat at Vaglaskogur on the north coast, together with redwings (*Turdus musicus*) and meadow pipits (*Motacilla pratensis*). Wren populations on smaller, isolated islands occupy even more spartan habitats, such as cliff faces on the Shetlands and St. Kilda, and heather moors on the Outer Hebrides. In these low, exposed habitats they encounter at most one or two potential competitor species, and sometimes none at all. Paralleling the decreasing luxuriance of these habitats is a decreasing clutch size which reaches lows of 5.0 eggs/clutch on the Shetlands and 4.9 eggs on St. Kilda (Figure 45).

It is reasonable to hypothesize that the decreased brood size in the island wrens is a response to decreased food supply. But available food should depend on several factors: insect

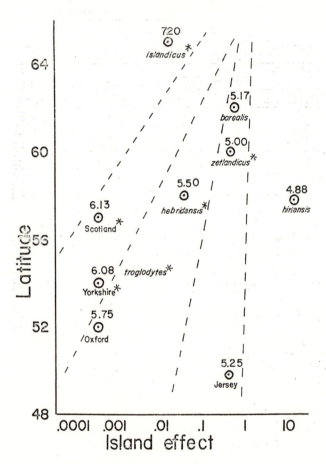

FIGURE 45. Clutch size variation in island and mainland populations of the wren *Troglodytes troglodytes* in the northeast Atlantic. "Island effect" is measured as the quotient (distance from mainland)/(island size). From Cody and Cody, 1972.

density, territory size, and competitor density. The third factor enters because some of the food potentially available to wrens is taken by competitors. Thus, territory size can be adjusted down to reflect the "removal" of these competitors, such that its inverse is wren-carrying capacity K. Adjusted territory size becomes $T = K^{-1} = (X_1 + X_2\alpha_{12} + X_3\alpha_{13} \ldots)^{-1}$, where X_1 is wren equilibrium density and X_2, X_3, . . . the equilibrium densities of its competitors, and the α_{1i} the competition coefficients between wrens and competitor species. For each wren pair in each location that the wrens were studied, we were able to compute a competitor-corrected territory size, and an insect density, measured with the aid of sticky plaques placed at various foraging spots, corrected for differences in the wren's preferred foraging heights in different habitats. Figure 46a shows that the fit is indeed rectangular hyperbolic (the model which provides the best least-squares fit is log $(T) = \log (F) - \log (I)$, with F a constant, I insect density, and T crude territory size); furthermore, the fit becomes considerably tighter when territory size is competitor-adjusted before regression against insect density, as this reduces the residual sums of squares by a factor of ten (Figure 46b). We can finally plot F, a measure of food availability, against clutch size, and, because only 20% of clutch size variability is explained by variation in food availability, the clutch size-food availability hypothesis is rejected, and other explanations must be sought.

C. Island Song Sparrow in the Pacific Northwest

Island wren population levels are lower than those of mainland wrens because, despite the low numbers of competitors on offshore islands, the food densities there are very much lower. In addition, the wren is not a particularly generalized species, i.e. it does not show high niche overlap values with its competitors, and therefore density compensation in the spe-

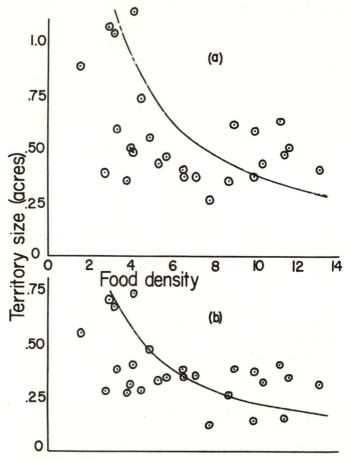

FIGURE 46. (a) Wren territory sizes are smaller where their insect food is more dense; (b) when the variable effects of competitors for insect food are removed, the relation becomes tighter.

cies is not well pronounced. Neither of these restrictions marks the song sparrow populations on islands around northwest Washington State studied by Yeaton and Cody (1973). This is because the song sparrow habitats on islands in Puget

Sound, the San Juan archipelago, and off the Olympic Peninsula are very similar to those on mainland Washington and
elsewhere in the United States. The Wyoming willows plot
discussed earlier covers habitat typical of song sparrows in
many areas of the country, including these offshore islands.

A plot of song sparrow territory size against the number
of competitors (the number of other passerines in the territory) is shown in Figure 47, for five mainland communities
of 9–14 species and 25 simple island communities of 1–9
species. The relation is linear; larger territories are presumably required where larger numbers of competitors reduce
song sparrow food density. The extremes are 2.24 acres per
territory in the Wyoming willows plot (11 competitors) and
0.02 acres on West Dock Island, Washington (no competitors).
Of course there are *some* differences in song sparrow habitats
from place to place, as shown by differences between foliage
profiles. It might be supposed that some of the decrease in
song sparrow territory size on the islands is due to increased
habitat complexity there with more food/unit area; but the
island vegetation, if different at all, is invariably less complex,
and lower. In spite of this, the vertical foraging range of the
song sparrow is expanded on the islands; in comparison with
song sparrows on "mainland" Vancouver Island at Saanichton, island birds on Mandarte, three miles offshore, spend
22% more of their time on the ground, in the absence of robin,
Swainson's thrush, California quail, and rufous-sided towhee,
and 11% more in the zone 6′–15′, in the absence on
Mandarte of Bewick's wren, orange-crowned warbler, yellow
warbler, common bushtit, and chestnut-backed chickadee.
Song sparrows also use grasslands on Mandarte, from which
they are presumably excluded on the mainland by savannah
sparrow, western meadowlark, skylark (*Alauda arvensis*), and
Brewer's blackbird. All of these competitor species have been
filtered out of the Mandarte community by the island's isolation and small size, and only rufous hummingbirds, barn swal-

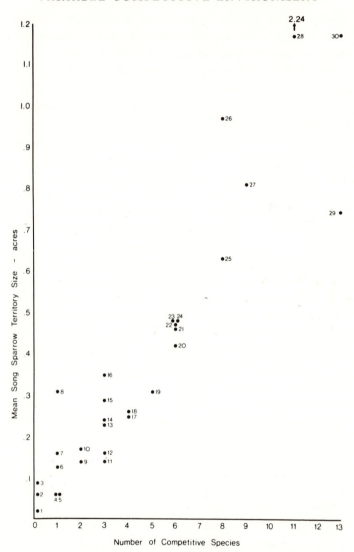

FIGURE 47. Song sparrow territory size as a function of the number of other passerine species with which it shares its territory. From Yeaton and Cody, 1973.

lows (*Hirundo rustica*), and red-winged blackbirds (*Agelaius phoeniceus*) remain. Food densities also vary between island and mainland, as they do between times of the day and between different days. However, insect abundance varied in no systematic way between island and mainland sites, and thus was left out of further analysis.

In the way which by now is familiar, carrying capacities for the song sparrow are calculated on the basis of community matrices and equilibrium densities in two mainland sites, Saanichton, Vancouver Island, and the Wyoming Willows. On some islands the species composition is not merely a subset of that in the mainland communities, but contains other species. In many cases these can be termed equivalent to mainland species (e.g. rufous hummingbird to Calliope hummingbird); in other cases the competition coefficients between song sparrows and the new competitors must be calculated.

The song sparrow data provide a unique opportunity to test an important aspect of competition theory. Three alternative combinations of components of competition coefficient (α_{ij}) can be tested for resultant goodness-of-fit between predicted and observed song sparrow density on the offshore islands. These are (a) "product alpha," $\Pi_k \alpha_{k,ij}$, the product of three overlap components $k = H, V$ and F, a measure of the volume (in three niche dimensions) which assumes no niche expansion and complete independence of niche dimensions; (b) summation or "expansion alpha," $\Sigma_k \alpha_{k,ij}/3$, the average niche overlap along the three niche dimensions, which deviates above direct proportionality with product alpha as the overlaps on the three dimensions deviate from equality and assumes interdependence of niche dimensions; (c) "partial expansion alpha," a measure in some sense intermediate between the first two, which presumes expansion into those parts of a competitor's niche most likely to be utilized in its absence. This is measured by simply extending the three faces of the niche overlap volume orthogonally throughout the competitor's niche, and measuring the resultant volume. Writing the volume of the original competitor-limited niche overlap as (α_k), expansion of this volume as described above becomes

$$\alpha_H\alpha_V \cdot 1 + \alpha_H\alpha_F \cdot 1 + \alpha_V\alpha_F \cdot 1 - 2\alpha_H\alpha_V\alpha_F$$
$$= \Pi(\alpha_k) \left[\frac{1}{\alpha_F} + \frac{1}{\alpha_H} + \frac{1}{\alpha_V} - 2 \right].$$

Figure 48 shows that, of these three alternatives, expansion alpha **a** generally makes the most accurate predictions, although partial expansion alpha comes in a close second. Product alpha does considerably poorer. This indicates that the properties of **a** which empirically make it a better estimator of niche overlap within and between communities of variable resource predictability contribute also to its superior predictability of expanded island densities. This scheme not only provides a way of distinguishing between the appropriateness of various alternative forms for the elements of the community matrix but also an objective examination of a competition theory that has never been tested in its expanded form. Although far more of these tests are necessary, the close fit between observed and predicted island densities is extremely encouraging.

Apparently the song sparrow is able to compensate for the absence of its competitors on small islands by extensive niche expansions. Remarkably enough, it is able to make use of resources vacated by birds as different as towhees, quail, warblers, and thrushes! In fact, its success in colonizing offshore islands and persisting there may be due to a large extent to this capacity for large niche expansion in this generalist species. If complete niche expansion were possible, new island equilibrium densities could be calculated from the same equa-

$$K_1 = X_1 + \alpha_{12}X_2 + \alpha_{13}X_3 \cdot \cdot \cdot,$$

tion but as competition coefficients are now unity, by definition, the α_{ij} represent conversion coefficients to competitor

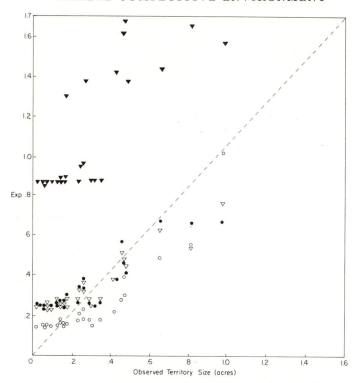

FIGURE 48. Observed territory size is plotted against territory size predicted for islands with various numbers of song sparrow competitors using as elements of the community matrix: i. product alpha **a** (solid triangles), ii. summation alpha **a** (open triangles), iii. partial expansion alpha (solid circles), and iv. complete expansion (open circles), which assumes that song sparrows can use vacant niches as well as could their true occupants if they were present.

equilibrium densities to reflect differences in body size and perhaps also differences in species-specific efficiency in converting similar foods to bird weight. Thus if niche expansion is nearly complete, total population density (of all species) on islands may well exceed, even by a factor of two or more, total population density on the mainland. This is especially true as larger species are underrepresented on islands, due both to their originally lower population densities and to their reduced ability to increase these following the drop-out of competitor species. The field studies by Grant (1966), Crowell (1962), MacArthur *et al.* (1972), and Yeaton (1972) that have documented higher overall population densities on islands indicate that these can be attributed to large density increases in just two or three species. These are presumably the small-sized generalists with wide capability for niche expansion.

III. SEASONAL CHANGES IN BIRD COMMUNITIES

Most temperate bird communities are composed in the breeding season of year-round residents plus summer visitors; the competitive environment of the residents can change in winter in a step-wise fashion as the summer visitors are replaced by winter visitors. Some of the available data are discussed in this section.

A. *Birds and Seasons in the Santa Monica Mountains*

1. The Habitats. The Santa Monica Mountains in southern California run fifty miles east-west across the northern side of Los Angeles basin and along the coast. The chaparral study site already mentioned is situated on the northern slope of these mountains at 2000′, and is located in tall and dense scrub vegetation, 6–10′ high, composed of diverse microphyllous and sclerophyllous plant species. Further down the north slope of the Santa Monicas toward the interior San Fernando Valley, especially in the moist canyons such as Cold Creek Canyon on the Murphy Ranch, 1750′, the chaparral runs into live oak (*Quercus agrifolia*) woodland. I censused 3.6 acres of this woodland in 1968 and 1969; it is 40–60′ high with a sparse understory of poison oak (*Rhus diversiloba*) and toyon (*Heteromeles arbutifolia*).

On the other slope of the Santa Monica Mountains the vegetation is very different. The effective precipitation is much less; this south-facing and ocean-facing slope is exposed to greater solar radiation, and the onshore winds further increase evaporation. The vegetation here is low and open and is dominated by the microphyll chamise (*Adenostoma fasciculatum*), 3–6' high. The climate in a 5.2 acre study site at the top of the south slope at 2220' is harsher and more variable, both within a daily cycle and over the seasons, than is the north slope chaparral, which in turn is more extreme than the climate in the oak woodland. Seasonal temperatures and weekly flying insect catches from these three study sites are given in Figure 49.

2. *Species Shifts with Season Along the Habitat Gradient.* Each habitat has summer visitors, the long-distance migrants which use the area for breeding only, year-round residents, and winter visitors. The two periods that best delimit the winter visitor-resident community from the summer visitor-resident community are mid-October to mid-February and mid-February to early June. Of the 44 passerines listed, 13 are summer visitors to the mountains, 20 are resident, and 11 are winter visitors. Two of the 20 residents in oak woodland are summer visitors to north slope chaparral, and four more are winter visitors there with two of these appearing in the same capacity in south slope chaparral. All but one of the resident species occur in the oak woodland; eight drop out in the north slope chaparral, and three more in the south slope chaparral.

The figures in Table 9 are proportions of the study areas occupied by species. This makes it possible to identify the centers of abundance (as territory size is not too variable between habitats) of each species. Thus species can be identified as preferring woodland (e.g. plain titmouse) or north slope chaparral (e.g. Bewick wren), but no residents reach maximal

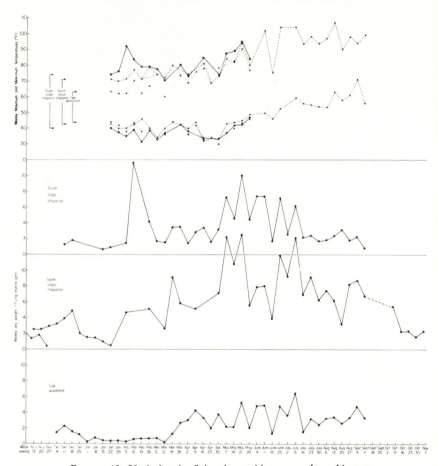

FIGURE 49. Variation in flying insect biomass and weekly temperature maxima and minima at three sites in the Santa Monica Mountains, southern California.

abundance in south slope chaparral. Only a few of the residents (e.g. Anna's hummingbird and plain titmouse) maintain constant densities between seasons in a habitat.

A common trend is for species to shift down the habitat gradient after breeding, toward a center of distribution closer

152

Table 9. Breeding and winter bird densities in three habitats of different resource predictabilities in the Santa Monica Mountains.

	Species:	Oak woodland		North slope chaparral		South slope chaparral	
	Habitats:	Br	Win	Br	Win	Br	Win
1	Bullock's oriole	0.10					
2	Black-throated gray warbler	0.51					
3	Black-chinned hummingbird	0.52					
4	Warbling vireo	0.90					
5	Western flycatcher	0.84		0.01			
6	Blue-gray gnatcher	0.66		0.05		0.06	
7	Ash-throated flycatcher	0.94		0.63		0.11	
8	White-throated swift	1.00		1.00		1.00	
9	Black-headed grosbeak	0.56		0.70		0.35	
10	Orange-crowned warbler	0.92	0.08	0.20			
11	Hutton's vireo	0.90	0.76	0.03			
12	House wren	0.97	0.13				
13	Song sparrow	0.90	0.27				
14	Canyon wren	0.27	0.34		0.22		
15	Swainson's thrush	0.34	0.20		0.52		
16	House finch	0.68	0.13		0.14		1.00
17	Lesser goldfinch	0.85	0.74		0.24		1.00
18	Nuttall's woodpecker	0.94	0.74	0.46	0.08		0.05
19	Plain titmouse	0.94	0.92	0.44	0.43		0.21
20	Red-shafted flicker	0.61	0.63	0.71	0.50		0.70
21	Anna's hummingbird	0.89	0.90	0.88	0.89	0.67	0.59
22	Spotted towhee	0.74	0.94	0.80	0.62	0.61	0.17
23	Bushtit	0.92	0.26	0.95	0.85	0.57	0.59
24	Scrub jay	0.68	0.84	0.92	0.96	0.70	0.57
25	Bewick's wren	0.10	0.58	0.92	0.83	0.49	0.57
26	Brown towhee	0.23	0.35	0.62	0.65	0.81	0.42
27	Wrentit	0.21	0.65	0.99	0.91	0.87	0.49
28	California quail	0.13	0.50	0.78	0.73	0.62	0.10
29	California thrasher		0.06	0.94	0.94	0.94	0.61
30	Costa's hummingbird					0.62	
31	Black-chinned sparrow					0.39	
32	Rufous-crowned sparrow					0.31	
33	Sage sparrow					0.30	
34	Golden-crowned kinglet		0.71				
35	Rock wren		0.10				
36	Audubon's warbler		0.60		0.31		
37	Hermit thrush		0.03		0.16		
38	Ruby-crowned kinglet		0.97		0.74		0.17
39	Oregon junco		0.53		0.43		0.66
40	Robin		0.11		0.10		0.10
41	Fox sparrow		0.11		0.82		0.08
42	Waxwing				0.10		0.10
43	Golden-crowned sparrow				0.24		0.85
44	White-crowned sparrow				0.07		0.85

TABLE 9. (*Continued*)

Species:	Habitats:	Oak woodland		North slope chaparral		South slope chaparral	
		Br	Win	Br	Win	Br	Win
(1) Number of species:		28	27	19	25	17	21
(2) % Visitors by species:		32	30	37	52	47	57
(3) % Visitors by individuals:		33	24	22	33	33	58
(4) Point diversity:		18.25	13.12	12.03	12.48	9.42	9.88
(5) % Seasonal change:		−28.7		+3.7		+4.9	
(6) Insectivore point	Visitors:	5.37	2.38	1.69	1.27	1.17	1.13
(7) diversity:	Total:	14.88	7.85	9.33	5.70	6.55	3.19
(8) % Insectivores:		81.6	59.8	77.6	45.7	69.6	32.3
(9) % Seed eaters:		18.4	40.2	22.4	54.3	30.4	67.7
(10) Seasonal	Insectivores:	−47.2		−39.9		−52.1	
(11) change in	Seed eaters:	+56.4		+197.4		+133.1	

to the more stable and predictable oak woodland. For instance, Bewick wren, wrentit, and California quail are only marginal breeders in oak woodland, but are common there in winter. In fact, quail are as marginal in winter in south slope chaparral as they are in the breeding season in oak woodland. Seven of the residents follow this pattern. The reverse trend occurs in several species, most notably in finches and in only one foliage insectivore, the bushtit. The partial exodus of south slope chaparral breeders in the winter leaves this habitat more open to invasion by winter visitors.

3. Changes in Community Composition with Season. The numbers of breeding season bird species bear the usual relation to habitat structure, and decrease from 28 in oak woodland through 19 to 17 in south slope chaparral. The numbers of wintering bird species, on the other hand, increase over the breeding season totals in all sites except the oak woodland; this indicates that the increased equitability and predictability of the woodland permit a persistence of community organization year-round and a resistance by this community to large numbers of winter invaders. Larger numbers of ecologically similar species pack into the other two habitats, which, according to their reduced predictability and diversity of resources, should be less able to support them. It seems that the ecological segregation patterns that characterize these communities in the breeding season do not persist through the nonbreeding season.

Table 9 shows how community composition changes with site and summarizes at the bottom the composition of each community. Line (1) (in Table summary) shows how species number remains constant year-round in the oaks, but increases up the habitat gradient. Line (2) indicates that residency decreases and species turnover between seasons increases up the gradient, and (3) shows that visiting species are each as common as residents in summer in the oaks and in winter in south slope chaparral, but elsewhere and at other seasons the residents easily outnumber the visitors. Line (4) gives the point diversity of each community, the average number of species whose activities encompass a point in the habitat selected at random and measured by simply summing over species the proportions of the habitat they each use. From breeding season to wintering season there is a large reduction in point diversity in the oaks (5), but winter point diversity exceeds breeding point diversity in the other two sites. The last four lines summarize the information obtained from broadly classifying species as insectivores or seed/fruit eaters. Community differences are illustrated in Figure 50.

The oak woodland can be characterized by small seasonal turnover, as around a third of its individuals and species, almost all insectivores, are replaced by an equal number of species (but a reduced number of individuals) of chiefly insectivorous winter visitors. The addition of one common winter visitor, Oregon junco, and changed food habits of some resident species (jays, towhees, wrentits) decreases the proportion

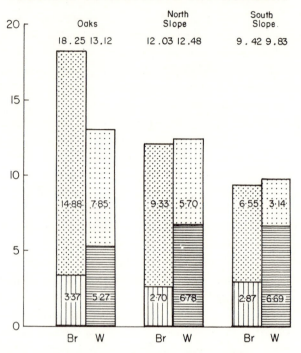

FIGURE 50. Bird species diversity in oak woodland and two chaparral sites in the Santa Monica Mountains of southern California. Diversity is measured as "point diversity," the number of species that include the average point in their foraging ranges or territories. Br = breeding season, W = wintering season; stipples are insectivores, hatchings are seed- and fruit-eaters.

of insectivores 22%. At the other extreme in south slope chaparral, the breeding population is again one-third summer visitors by individuals, but is almost half visitors by species. In winter, likewise, visitors are over twice as predominant as they are in woodland, and the many ecologically similar finches that winter here reduce insectivore predominance 37%, to just one-third of the total individuals present (vs. almost 2/3 in woodland).

4. *Ecological Overlaps in Breeding and Wintering Communities.* We have already learned that interspecific habitat overlaps are sensitive indices of the resources and competition environment; habitat overlaps for residents and visitors in the Santa Monica Mountains are given in Table 10. Habitat over-

TABLE 10. Habitat overlaps among visitors and among residents in breeding and wintering communities in the Santa Monica Mountains.

	Oak woodland		North slope chaparral		South slope chaparral	
	Br	Win	Br	Win	Br	Win
Residents X residents	0.662	0.584	0.782	0.720	0.713	0.722
Visitors X visitors	0.599	0.307	0.608	0.344	0.417	0.689

laps among the resident species are uniformly high, being within the range 0.6–0.8 in all but one case, irrespective of the habitat and the season. This indicates that the resident species, which compose their own constant competitive environment year-round, maintain differences from each other in food and feeding behavior, and are habitat generalists. They are apparently little affected by the seasonal turnover in the visitor segment of the community.

Looking now at the visiting species, we find that those that arrive to breed in oak woodland and north slope chaparral are insectivores (warblers and vireos), which maintain the high habitat overlap typical of food specialists. Presumably the increased insect spectrum in the spring opens up new niche opportunities to these species, which divide food resources as discretely as do the residents. On the other hand, the habitat-use patterns of the summer visitors to south slope chaparral are more typical of opportunistic food generalists, species which differ from each other chiefly in where they live, not in what they eat. These summer visitors are mostly finches that have strong bills capable of handling a variety of foods, and are known to be omnivorous in the breeding season. In fact, three small finches that show up to breed in south slope chaparral tolerate less than 25% habitat overlap among themselves.

Winter visitors to oak woodland and north slope chaparral have much reduced habitat overlaps. They spend their time in select parts of the habitat, a behavior we expect from food generalists coexisting in the same habitat. Division of habitats rather than food or feeding locations is particularly apt if these species are surviving on the less predictable food supplies of the two habitats. But in south slope chaparral, even though the wintering species there are again mostly finches that can potentially behave as food generalists, habitat overlaps are much higher, at a level typical of resident species. The winter visitors to south slope chaparral make no attempt to subdivide habitat, but rather have evolved interspecific social behavior and spend much of their time in mixed species flocks. The heavy finch concentrations feed on *Adenostoma* and *Salvia* seed crops, and also on the seeds of annuals, which are more abundant in the open south slope habitat. Within these mixed-species aggregations, species appear ecologically indistinguishable. Apparently the diversity, abundance, and dispersion of these seeds preclude an orderly food allocation either within or between species; the remaining alternative is ecological convergence and food exploitation by mixed-species groups.

These results show many similarities to the results of Emlen's (1972) study of winter bird communities in Texas. He contrasted a variety of habitats from grassland to river forest, and although the ratio of wintering to resident species was highest in the river forest, the influx of wintering individuals was highest in the more open habitats, where flocking is common. In addition, habitat overlaps were highest overall in the climatically and presumably resource-buffered forest habitats, and were higher among resident species than among the wintering birds. Fretwell (1972) has discussed the evolution of winter habitat use in a more complete context of food availability and finch survivorship in various habitat types and relates this to breeding habitat preference; his valuable book should be consulted by all students of seasonality.

B. Seasonality in the Mohave Desert

The Santa Monica Mountains enjoy mild and predictable winter climates. Although not enough is known of seasonal changes in the resources there, apparently birds find food sufficiently predictable and diverse year-round to maintain structured communities of ecologically discrete species both winter and summer in all habitats except the more seasonal south slope chaparral.

In the Mohave Desert, however, there is a much more drastic change in resource spectrum with seasons. The breeding birds (Appendix A) number twelve, and although some adults may eat seeds and fruit (house finch, black-throated sparrow, phainopepla), almost all feed insects to their young. Winters in the Mohave Desert are harsh and unpredictable. It frequently snows (at 4000'), cold winds sweep the area, and insects are buried in the ground or under bark and are available to only thrashers and verdins.

Yet winter birds are conspicuous and common in the Mohave Desert; they are very predominantly seedeaters. Quail, the black-throated sparrow, and house finches are resident, but their numbers are greatly increased by many other finches. Sage sparrows, white-crowned sparrows, golden-crowned sparrows, and Oregon juncos are common winter visitors, and from late February to April passage migrants are present. In order of abundance, these are Brewer's sparrow, chipping sparrow, black-chinned sparrow, green-tailed towhee, and black-headed grosbeak. All are seedeaters, yet the diversity of seeds is low. The three seed types which collectively comprise almost 100% of the finch food are *Erodium,* various mustards

(Cruciferae), and, after good rains, the grasses *Schizmus* and *Bromus* (Cody, 1971).

The wintering finches spend almost all of their time in flocks. This desert has some late summer rain but gets mostly winter storms, and if rainfall has been at or above the long-term average of around 6″, the flocks are small and monospecific. If rainfall has been low, however, the flocks become very large (500+ individuals) and contain most of the species mentioned above. Within these large flocks there appear to be two or three ecological roles, distinguished by whether species take seeds from the annuals a) before they are shed, b) from the surface of the ground, or c) by scratching, after the seeds have been blown around for a while on the ground and are partly hidden. The number of species in the flocks far exceeds the number of distinct roles each might perform. The coexistence and survival of these species seems to depend on the formation of the flocks, in which they use similar behaviors to find the same foods. The evolution of contact notes that are convergently similar between species and the coherence and organization of these flocks attest to their selective advantage.

I have hypothesized that such flocks of wintering species have evolved to exploit more efficiently a food supply which is both sparse and finely dispersed (Cody, 1971, 1973b). The flocks have movement characteristics such as speed, frequency of turns, and probable direction of turns. These vary with flock composition, and effectively regulate the mean and variance of the time interval between successive visits to the same point on the desert floor. More rapidly renewing seed supplies permit more rapid returns to a feeding station, and this happens in drier parts of the desert where flocks move faster, turn more often, and turn more sharply. When food abundance is low, on the other hand, as it is where effective moisture is low, the flock must move faster if each individual feeds at the same rate. This will again promote reduced return time. Seed production and ripening rates in the Mohave Desert are chiefly functions of runoff, so that effective moisture is high close to the base of the mountains, and decreases with distance out onto the desert floor. The variables involved with finch flock economics are summarized in Figure 51; this represents an

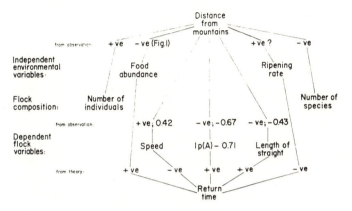

FIGURE 51. Environmental variables which affect the size, composition and behavior of flocks of wintering finches in the Mohave Desert. From Cody, 1971.

alternative food exploitation scheme that is favored over an individual- or species-specific allocation scheme. The phenomenon of flocking in nonbreeding finches is of course widespread and is presumably an adaptation to food supplies of low diversity, low density, and high dispersion.

CHAPTER FIVE

Parallel And Convergent Evolution

In this chapter I will examine the extent to which bird niches show similarities in response to parallel selective forces. The problem can be approached at various levels, for parallel evolution can be demonstrated within a broad range of sample sizes or units on each of two variables. The first axis is a gradient of decreasing proximity of habitat sites, so that comparisons may be made initially within habitats, then between adjacent habitats, and extended to comparisons on a continental and finally an intercontinental scale. A second axis ranks various sample sizes of numbers of bird species used in these comparisons. The comparisons may be made at the level of the single species, of groups of related or coexisting species, of whole communitites or of the complete avifauna. Here I will emphasize the comparison of communities between continents. At these levels parallel/convergent evolution is at its most dramatic, for the former choice extends the matching of individual species niches to include the relative juxtaposition of these niches in the community, their number and shape, and the latter condition assures us that the species in these communities so compared will bear minimal taxonomic affiliation to each other and thus minimal resemblance due to factors other than parallel selection. While comparisons at different levels of species organization and geographic separation will be treated in turn as available information allows, intercontinental community comparisons between North and South America will be given special attention.

PARALLEL AND CONVERGENT EVOLUTION

I. CONVERGENCE AT THE LEVEL OF THE SPECIES

A. *Replacement of Species on a Local Scale*

1. Within-Habitat Replacements. First mention will be given to the few instances in which species replacements within habitats show apparent effects of convergent evolution. However, such cases are speculative, as most such replacement pairs are closely related, and we seek to explain not so much a convergence in appearance and behavior as an exceptional lack of divergence. The two sibling species of meadowlarks *Sturnella magna* and *S. neglecta* (Figure 1) show little or no divergence in feeding behavior and morphology. This lack of divergence is in a sense tantamount to parallel evolution, especially when the same appearance and behavior have been evolved in unrelated grassland species elsewhere (see below; the meadowlark niche will be a *leitmotiv* of this chapter). Likewise the grasshopper sparrow *Ammodramus savannarum* is replaced by its congener Baird's sparrow, *A. bairdii,* in the northwestern part of its range (Figure 1c). The two are virtually identical in ecology and behavior and very similar in appearance (as are many grassland sparrows). These cases are further complicated, beyond the taxonomic factor, by the fact that these two species pairs are interspecifically territorial where they meet. Thus there are three selective forces promoting similarity in appearance: a) common ancestry, b) common niche characteristics, and c) selection for a common signal (appearance and/or voice) used in territory defense. In other finch genera such as *Junco, Aimophila, Amphispiza,* and *Pipilo,* the same factors are probably operating, and likewise in the Chilean finch species of *Spinus* and *Phrygilus.*

Discussion below will be restricted to species pairs in which the second factor in phenotype convergence, that of niche coincidence or near-coincidence, can be singled out. But it should first be stressed that there are many taxonomically related species pairs and groups which fail to show divergence in appearance although they have become reproductively isolated (including the so-called sibling species). The genera *Empidonax* and *Myiarchus* are represented in North America usually by not more than one species in any one habitat. European warblers such as *Sylvia* and *Phylloscopus* provide similar examples. Within each of these genera the species pose problems in field identification, and, in the first two genera at least, species may be interspecifically territorial; the problem of isolating the various selective forces complicates such examples.

2. Species Replacements Between Habitats. The phenomenon of parallel or even convergent evolution can be seen locally in the replacement of species among habitats in the same geographic area. Consider again the thrashers (Figure 2). Although classified into several species and two genera, the thrashers are similar in body size and build with sturdy legs and long tails; with the exception of two species which occupy drier country with perhaps harder ground (*Oreoscoptes montanus* and *Toxostoma bendirei*), all possess long, decurved bills to dig for their food. None occurs in forests, but there is a different species for each of several different types of brushy habitats. Apparently the adaptations to the thrasher feeding niche are common to several variations of scrub habitat, and largely override the habitat-specific selection to which each species is exposed. Woodpeckers (Aves: Picidae) and hummingbirds (Aves: Trochilidae), among many other possibilities, provide similar examples of species groups in which a turnover in species names but not in species ecological role occurs with a change in habitat. We might say of such examples that the niche is not confined to a single habitat type, but remains relatively unaltered with minor or even considerable changes in habitat structure and geographic location.

B. Intercontinental Species Replacements

1. Historical Aspects. When a turnover among ecologically equivalent species takes place between taxonomically unrelated species in habitats which themselves are floristically unrelated and geographically isolated, dramatic convergences can result. Here we are convinced that the component of natural selection which is comprised by habitat is dominant in producing the convergence, and that genealogy and history play minor, usually conflicting, roles.

The first indication that species on different continents could become very similar in appearance and gross morphology in response to selection for parallel niches came from the descriptions of the old systematists. When biological exploration took place in new worlds, the material was received by taxonomists whose basis of comparison was their own familiar fauna. What could be more natural and, indeed, what a fine ex tempore test of the convergent evolution concept, than that they should name the new forms, on the basis of gross morphology, after the closest local types? Of course, it is just this gross morphology which so readily becomes adapted by natural selection to a particular niche, and structurally similar habitats provide similar templates for specific niches. For the sake of science and to the detriment, perhaps, of an ecologically oriented traveler, the true taxonomic affiliations of these exotic species were eventually revealed; feathers were counted, bones numbered, and muscles displayed, and new and different families and genera were discovered to have come to resemble and replace in distant lands the local fauna. Thus Linnaeus (1758) first classified the North American warblers after the Old World warblers (Sylviidae) he knew from Sweden rather than as a distinct family (Parulidae). Likewise the North American mimic thrushes Mimidae were classified with the Old World thrushes Turdidae, its vultures (Carthartidae) with the Old World vultures (Aegypiidae), and its blackbirds (Icteridae) with the Old World blackbirds (also Turdidae). In this way Linnaeus inadvertently and unknowingly came across one of the great cases of convergent evolution. A species from the North American prairies was classified as a lark *Alauda magna* (Alaudidae) by Catesby, and Linnaeus concurred. It is found, said Linnaeus, not only in America but also in Africa! Buffon (1812) was skeptical of such a range. In fact two species were involved, neither is a lark, and neither is closely related to the other. The American bird is the meadowlark *Sturnella magna* (Icteridaè), and the other is the African pipit *Macronyx croceus* (Motacillidae), whose similarity to *Sturnella* is now legendary (see Friedmann, 1946).

Similar confusion, albeit entertaining and even educational confusion, reigned when the systematists began naming South American birds. To one familiar with the habits and appearance of European and New World species, all the "right" mistakes were made. In Chile, for example, many of the small passerine niches are occupied by flycatchers Tyrannidae and ovenbirds Furnariidae, families restricted to the New World and dominant in South America. Many tyrannids were named *Muscicapa* after the Old World flycatchers Muscicapidae, open country forms of both families (*Lessonia, Geositta, Cinclodes*) named *Alauda, Motacilla,* or *Anthus* after common open country birds in the lark and pipit families in the Old World, and foliage-hunting forms (*Tachuris, Phleocryptes*) called *Sylvia* after the Old World warblers. The designation of *Eugralla* (in the endemic South American family Rhinocryptidae) as *Troglodytes* (a wren) is particularly apt. These examples could be multiplied at great length, but the point has been made. Works calling attention to these overall similarities between intercontinental species and familial replacements are numerous and include Friedman (1946), Moreau

(1966), Lack (1968, 1969), Fry (1970), and Vuilleumier (1971), as well as the South American bird guidebooks.

2. *Species of Simple Habitats.* Many examples of convergent evolution may be drawn from species which occupy simple habitats. This is because a) such habitats may be compared and matched with greater accuracy, b) numbers of species, and hence the possibilities for niche rearrangements and alternative juxtapositions, are few. Grasslands, deserts, and tundra are examples of habitats which from the avian ecology viewpoint may be regarded as simple. Many of the classic cases of convergent evolution are found in these habitats: the mammalian examples of *Dipodomys* (Heteromyidae, North American deserts), *Antechinomys* (a marsupial in Australian deserts), *Gerbillus* and various Dipodidae (Asian and African deserts); the reptile example of *Phrynosoma-Moloch* (North American and Australian deserts, respectively); the grassland birds *Sturnella-Macronyx,* plus their South American replacements *Pezites* and *Leistes* (both Icteridae). This bird complex is quite remarkable, for all species are terrestrial and walk around in grasslands using a pickaxe-like bill to dig and poke for insects. The *Sturnella* (2 species) are streaked brown above, yellow below with a black breastband, a pattern exactly duplicated by the pipit *Macronyx croceus.* The South American *Pezites* (2 species) repeat this, except that the yellow is replaced by red. Curiously enough, a more southerly African species, *Macronyx ameliae,* is also red below rather than yellow; this may simply be part of the association of a red coloration with wetter and yellow with drier habitat which is general in birds. All species are alike in size, show similarities in voice, place nests in similar situations, and lay three to four eggs at latitude 30°. (*Pezites* also has an enlightening taxonomic history, for Linnaeus called it a *Sturnus,* later changed this to *Pezites,* which in turn became *Sturnella* (Filippi, 1847),

167

returned to *Pezites,* then back to *Sturnella* by Short in 1968.)

Marshes qualify as simple habitats for birds. In western North America their characteristic resident birds are the icterids *Agelaius phoeniceus, A. tricolor,* and *Xanthocephala xanthocephala,* the red-winged, tricolored, and yellow-headed blackbirds. The first two species are all black, with, respectively, buff-bordered and white-bordered red shoulder patches ("epaulets"). In Chilean marshes lives a single icterid, the yellow-shouldered marsh bird *A. thilia,* in which the epaulets are yellow rather than red; this color covers the head of the third North American species, *Xanthocephala.* As discussed by Lack (1968), the weaverbirds Ploceidae fill this niche in African marshes; some species (e.g. *Diatropura*) closely duplicate in appearance, behavior, and ecology these icterid species. Other marsh-dwelling passerines in North America include two wren species (Troglodytidae) and several Eurberizidae (*Ammospiza, Melospiza*). In Chile one of the same two wrens occurs, together with a multicolored tyrannid *Tachuris* and a sylviid-like ovenbird *Phleocryptes.*

The chats *Oenanthe* and *Saxicola* (*Aves:* Turdidae) are palaearctic and Ethiopian in distribution; a single species *Oenanthe oenanthe* has crossed the Bering Sea and breeds across northern North America while migrating southwest back to the Old World. The chats are characteristic of open habitats, deserts, stony heaths and commons, tundra and barren hillsides. They use elevated perches on rocks and pounce on flies and other moving insects. In South America the aptly named genus *Muscisaxicola* (Tyrannidae) provides for north temperate bird watchers extremely convincing "chats," in feeding behavior, habitat, preference and in breeding biology. The feeding behaviors of the wheatear *O. oenanthe* and *Muscisaxocola* species which bracket it in body size are compared in Figure 18. Several *Muscisaxicola* spp. in central Chile are interspecifically territorial (see below, Chapter 6); this also corresponds to what has been observed in chats, for Stresemann (1950) described the nonoverlapping territories of three wintering *Oenanthe* species in Egypt.

Another north temperate group which does not reach South America is the thrashers. In Chile and elsewhere in South America, the furnariids *Upucerthia* occupy the thrasher niche, and have come to resemble thrashers in many ways. They are large terrestrial birds of open and scrub country, and are equipped with long thrasher-like decurved bills. I am familiar with *Upucerthia dumetaria*, which in habitat most closely resembles *T. lecontei* or *T. dorsale*. The color plates in Peterson (1961) and Johnson *et al.* (1957) may be compared, and similarities in body proportions verified from Table 11.

TABLE 11. Body measurements (mm) in two north temperate thrashers (Mimidae) and the two south temperate ovenbirds (Furnariidae) which are their closest ecological counterparts.

Species	Body size	Tail	Wing	Bill	Range
Upucerthia dumetaria	210	84	96	30.7	Chile
Upucerthia vallidirostris	210	76	88	34.5	Argentina
Toxostoma lecontei	230	122	98	32.8	S.W. United States
Toxostoma redivivum	250	130	103	36.3	W. United States

A niche which must be present and constant over those parts of the world that are densely populated by man is that associated with his urban buildings and refuse, the "house sparrow niche." In Europe, and in North America since their introduction to New York in 1850's and subsequent spread to virtually all parts of the country, house sparrows (*Passer domesticus;* fam. Ploceidae) are a sight familiar to everyone. They must be described as "generalists," and perhaps a part of their success story (from humble, probably African, origins, to cosmopolitan ubiquity) is their willingness to nest almost anywhere (buildings, tree hollows or outer branches, pylons, rocks) and to eat almost anything (insects, fruit, seeds, garbage, feces). In California they have apparently displaced house finches (*Carpodacus mexicanus*) from heavily urban centers, and coexist with the finches in suburbs where there is sufficient greenery for the latter to survive. The native "house sparrow" of much of Central and South America is the rufous-naped sparrow, *Zonotrichia capensis,* whose extraordinary geographic range extends from Mexico to Tierra del Fuego. House sparrows were first seen in Chile in 1904 and have since colonized the whole country. In city centers the native chincoles, *Z. capensis,* are rarely seen, but as in the house sparrow-house finch situation, they still occupy gardens and city parks where they can apparently hold their own. In downtown Lima, Peru, I saw no *Zonotrichia,* but the plaza at Arequipa was about 50:50 *Passer-Zonotrichia,* and "downtown" Cuzco was occupied solely by *Zonotrichia* in 1965. House sparrows are evidently without competitive superiors in urban centers, but coexistence by some resource division is achieved at some point along an urban-suburban-rural gradient with the tree sparrow *Passer montanus* in Britain, the house finch in California and the rufous-naped sparrow in South America. As already mentioned, the order house sparrow to tree sparrow from urban to rural habitats is reversed in Sardinia (Figure 44), where it appears that eventually *P. montanus* will occupy and exclude *P. hispaniolensis* from the urban section of the gradient. This is a fascinating, ongoing process, and merits close attention.

II. CONVERGENCE AT THE LEVEL OF THE COMMUNITY

A. Habitats of Simple Structure

Many striking examples of parallel communities exist among structurally simple habitats. Patrick (1961) early emphasized that the number of species of producers may be very similar among simple habitat types. Working with fresh-water diatomaceous algae, she demonstrated that the rivers of northeastern North America supported about the same number of species per substrate area as rivers in the southeast; even tributaries of the Amazon in Peru provided a close match in numbers of species, though their names are quite different (Patrick 1964). Intertidal communities, especially those of sandy beaches, show close parallels in species number and organization among coasts of different oceans. The facts that many of the animal genera are cosmopolitan in distribution (e.g. *Donax, Emerita*) and that no precise area-controlled censuses have been made lessens the value of this example. The work on sea-bottom communities, developed by Thorson (1957, 1958, 1960) and to which Kuznetsov (1970) is a recent contributor, has been approached chiefly from the point of view of parallel community studies. The results illustrate, perhaps more convincingly than any other data, that, given a habitat (sea-bottom) simple and relatively constant in structure over large geographic areas, community structure in terms of numbers and types of species also remains constant. Some infaunal communities, such as the shallow water mixed-bottom *Macoma* associations, show quite precise parallels in the numbers of species and in their niche relationships among coasts as dissociate as those of Greenland, Denmark, Washington (northwest United States), and Japan.

Terrestrial communities of almost the simplicity of sandy beaches are provided by dry, sandy deserts. In such deserts remarkable parallels at both the level of the species and of the community may be found. The parallels between the lizard species of the Australian and North American deserts have been described by Pianka (1969a, 1969b, 1970, 1971), and Sage (MS) discusses those he finds between North and South America. Another species group intimately associated with the substrate in deserts is the ants. Kuznezov (1953, 1956)

171

has discussed parallelism as a general phenomenon and with particular reference to desert ants. Despite wide differences in taxonomic origins, the niches of "hunting ants" and of "graminivorous ants" show similar occupancy and adpatations between Central Asia (genera *Cataglyphis* and *Messor* respectively) and Argentina (genera *Dorymyrmex* and *Pogonomyrmex*). But a third niche group, fungus-farming ants, occurs only in the New World. In the Mohave Desert of southeastern California eight to twelve species of ants may be found in a three-acre patch of uniform desert at 4000′ elevation (Cody and Bernstein, unpublished); this number dimishes steadily to six to eight species at 2000′ (Bernstein, 1971). In Utah, one acre of substructurally similar Great Basin Desert supports an average 12 species of ants (communities 2, 4, 5, 6, 7 of Ingham, 1963), whereas an acre of Mohave Desert supports an average six species (communities 10–17, Ingham, *op. cit.*). In the central Sahara Desert, Bernard (1961) finds that 14 species regularly occur in the area of Tassili n'Ajjer. Further, the relative abundances of these ant species show similarities among desert locations, as do their size distributions.

Bird species are more intimately associated with the vegetation which grows in these deserts, and this varies much more in structure than the substrate on which it grows. Thus the Great Basin Desert vegetation grows to a height of about 1 m, the Mohave up to 2 m, and the Lower Sonoran vegetation up to 4-5 m; correspondingly the small-bird populations, with counts of 6, 13, and 17 species/ten acres, respectively, show less parallelism than lizards and ants.

B. *Marine Birds*

The habitat exploited by marine birds might be expected to show little variation from coast to coast. Six species of the bird family Alcidae breed on the Olympic Peninsula in the northwest corner of the United States (Cody, 1973a). Five of these species dive for fish, and can be ranked according to the distances from the breeding rock at which they feed. The sixth feeds on plankton at greater distances offshore.

Grimsey Island, northern Iceland, also supports six alcid species, five of which are different species from the Pacific set. A strong correspondence exists, however, in the organization of their feeding zones (Figure 52). The distances over which the birds have to travel for food is the predominant influence on morphological, ecological, and behavioral traits, and hence the correspondence between pairs of species which feed at the same points in the feeding zone sequence indicated extends to these additional characters. In each set of species the sequence of body sizes and bill morphologies is similiar. Only the species feeding closest to the breeding cliffs rear two young, the others but one. Chick growth rates and precocity are in both sequences inversely and directly, respectively, parallel to the distance the parents travel for food. These trends are summarized in Table 12.

The Alcidae are restricted in their geographic distribution to the northern hemisphere. In the southern hemisphere, and in particular in Chilean waters, their closest ecological counterparts are the diving petrels Pelecanoididae and the penguins Spheniscidae. The former, in particular, bear an extremely close resemblance in their morphology, ecology, and behavior to certain of the smaller alcids, at least as far as existing information goes. The morphology of the common Diving Petrel *Pelecanoides urinatrix* has been investigated in detail by Kuroda (1967), who finds striking parallels in proportions, osteology, and musculature between it and the slightly larger Ancient Auklet *Synthliboramphus antiquus*. Unfortunately for a United States-Iceland-Chile comparison, almost nothing is known of the habits and breeding biology of the southern birds. The only life history study is that of Richdale (1943) on *P. urinatrix*. In common with the Ancient murrelet and with most of the smaller murrelets and auklets, *urinatrix* is a burrow nester, rears but one chick which is visited at night

NORTH ATLANTIC

NORTH PACIFIC

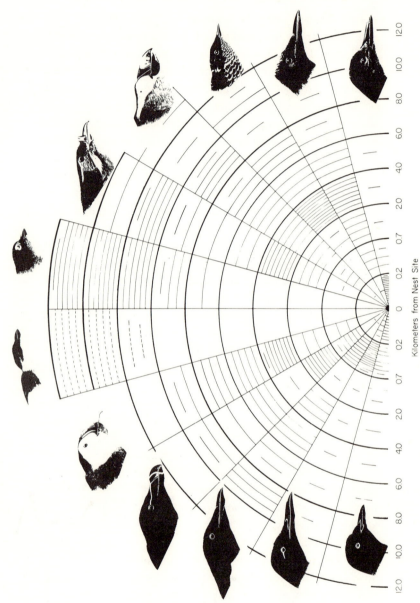

Kilometers from Nest Site

Figure 52. See legend on opposite page

and fed small fish and the larger crustaceans of the zooplankton, and is molested by the local larid (gull) predator. Judging by the chick growth rate of 5 gm/day and its attainment of a maximum weight in excess of that of the adult, the parents feed at considerable distances offshore.

The penguins Spheniscidae also show many characteristics which are convergently similar to those of alcids. Indeed, after the Alcidae was grouped with Colymbidae (divers, loons), during the period 1840–1893, they were next classified with Spheniscidae; the extinct great auk still bears the name *Pinguinus impennis*. Gysels and Rabaey (1964) discuss the systematic position of the Alcidae as evidenced by an electrophoretic study of lens proteins. Several genera showed divergent patterns, and the murre *Uria aalge* in particular appeared to be similar to the Spheniscidae with respect to this character. These authors cite figures which indicate that the Alcidae and the Spheniscidae hold 54% anatomical characters in common, while another 27% are variable between the groups, and conclude that the families show common origin and parallel evolution.

In spite of the evident morphological and adaptive similarities between the Alcidae and southern hemisphere replacements, the possibility of ordering a set of southern species into a feeding zone sequence as above does not look good. In the absence of the most appropriate data, Murphy (1936) is perhaps the best source of information. It appears that on most of the Chilean coastline only one *Pelecanoides* and one *Spheniscus* coexist. In the extreme south perhaps two of the

FIGURE 52. Distributions of foraging distances from island nest sites in two six-species alcid communities, Grimsey Island, Iceland, and Carroll Island, Washington State. Species are (clockwise from lower left) black guillemot, common murre, Brünnich's murre, razorbilled auk, Atlantic puffin, little auk, Cassin's auklet, rhinoceros auklet, tufted puffin, marbled murrelet, common murre and pigeon guillemot. Each complete arc represents 5% of total fishing time. From Cody, 1973a.

TABLE 12. Feeding and breeding biology of two six-species communities of marine birds (Alcidae), in the northeastern Pacific Ocean (Olympic Peninsula) and north Atlantic Ocean[1]

	Species	Body size (inches)	Distance to food (km)	Time in nest (days)	Chick growth rate (gm/day)	Precocity[2]	Nest site	Daily rhythm
North Pacific	*Cepphus columba*	14	0.27	35	10.7	9/10	crevice	diurnal
	Uria aalge	16½	3.10	21	8.5	1/3	ledge	diurnal
	Lunda cirrhata	14½	4.67	55	7.7	3/4	burrow	diurnal
	Brachyramphus marmoratus	9	?	(42?)	(5–6)	(4/5)	(burrow)	nocturnal
	Cerorhinca monocerata	13½	5.34	61	5.0	4/5	burrow	crepuscular
	Ptychorhamphus aleuticus	8	20 ±	45	3.5	7/6	burrow	nocturnal
North Atlantic	*Cepphus grylle*	13	0.47	35	9.1	7/8	crevice	diurnal
	Uria aalge	16½	0.59	23	6–9	1/3	ledge	diurnal
	Uria lomvia	18	3.28	23	6–9	1/3	ledge	diurnal
	Alca torda	17	2.62	17	6.8	±1/4	niche	diurnal
	Fratercula arctica	12½	3.97	42	5.0	10/13	burrow	diurnal
	Plautus alle	8	10	?	?	?	burrow	(nocturnal)

[1] From Cody, 1973a.
[2] Maximum proportion of adult weight gained in nest.

latter co-occur, and R. Schallenberger (pers. comm.) has taken two *Peleanoides* species together in the canals, but this is still far from a six-species group. Apparently the vacancies are filled by numerous Procellariformes, but, pending field studies, no further conclusions can be drawn.

C. Grasslands

Grasslands are simple habitats which can be duplicated in structure with high fidelity on different continents. Different families of birds predominate in different regions. In North America these are Icteridae and Emberizidae, with very few Motacillidae and Alaudidae. In South America the grassland passerines belong to Emberizidae, Furnariidae, and Tyrannidae, with an occasional Motacillid. In Asia, Fringillids, Motacillidae, and Sturnidae are encountered, and in Africa there are many grassland Alaudidae and Sylviidae.

Many grasslands, including those of widely scattered north and south temperate zones, the neotropics, and even arctic and antarctic heathlands, show striking parallels in bird species number and composition. In homogeneous areas around 10 acres in size, three or four passerines are present; a large vegetarian "grouse-type" species, both long- and short-billed "wader" species, and two or three raptorial species (including a "pursuer," often *Falco,* and a "searcher," often *Buteo*) are usual (Cody, 1966). Other grassland bird censuses confirm this picture (Alm *et al.*, 1966; Kuroda, 1968; Nakamura, 1963; Nakamura *et al.*, 1968; Ogasawara, 1966; Soikkeli, 1965; Williamson, 1967), and those which on the surface appear to contradict it, such as the South African grasslands and their avifauna, have not yet been censused at the 10-acre resolution required to test the apparent discrepancies (larger areas, with greater habitat heterogeneity can produce higher species counts, Figure 5).

Similarities between Chilean and North American grassland bird communities extend much further than parallels between numbers of species. A mixed grassland of intermediate height in Kansas supports grasshopper sparrows (*Ammodramus savannarum*) in the tallest and densest vegatation (mean = 1.02 ft), horned larks (*Eremophila alpestris*) in the shortest, open areas (0.76 ft), and eastern meadowlark (*Sturnella magna*) in intermediate vegetation (0.96 ft) with considerable overlap between adjacent pairs (Figure 6). In a field of comparable structure in central Chile, one also finds three passerines. *Pezites militaris* duplicates the meadowlark niche in those sections of the field intermediate in grass height

TABLE 13. Comparison of grassland bird habitats and phenotypes between Kansas and Chile (from Cody, 1973b).

Field characteristics

	# passerines	Mean Veg. Ht.	Vertical density	Profile area	Horiz. density
Chile	3	0.89'	6.23	39.22	9.26
Kansas	3	0.97'	5.10	32.35	8.68

Bird habitat preferences

Chile				
Pezites militaris	0.86	6.07	36.19	8.81
Sicalis luteola	0.93	6.62	43.82	9.86
Anthus correndera	0.86	5.90	36.25	8.91
Kansas				
Sturnella magna	0.96	5.12	32.81	8.87
Ammodramus savannarum	1.02	5.33	34.15	8.76
Eremophila alpestris	0.76	4.06	22.97	7.56

Bird morphologies[1]

	Body length	Wing	Tail	Bill length	Bill depth	Bill ratio	Tarsus
Chile							
Pezites militaris	264	117	98	33.3	13.3	0.40	34
Sicalis luteola	125	70	49	9.7	7.1	0.73	16
Anthus correndera	153	79	60	13.0	5.5	0.42	23
Kansas							
Sturnella magna	236	122	79	32.1	11.6	0.36	42
Ammodramus savannarum	118	62	47	10.8	6.5	0.60	19
Eremophila alpestris	157	104	69	11.2	5.6	0.50	22

[1] All measurements in mm.

and density (Table 13) and displays a corresponding feeding behavior (Figure 53). *Sicalis luteola* occurs only in the tallest parts of the field, and therefore replaces the grasshopper sparrow. In the most open parts of the field, lives *Anthus correndera,* doing just what the horned lark does in Kansas. These matches in feeding behaviors are remarkably close (Figure 53), as are those of the species' body proportions on which the behavior is dependent (Table 13).

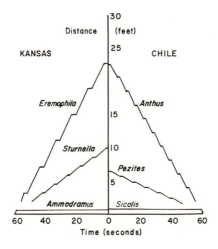

FIGURE 53. Comparison of feeding behaviors of the three bird species resident in each of two fields of medium grass height, one in Kansas and the other in central Chile.

Thus the Chile-Kansas convergence extends from number of species (and their relative abundances) to the precise way in which habitats are subdivided and food harvested, and the resultant behavior and morphological characters. The net outcome is that the ecological overlaps among species pairs within each study area are very similar; mean (pair-wise) habitat overlap is 63% in Kansas and 60% in Chile, the extent of vertical overlap in feeding heights (78% in Kansas vs. 89% in Chile) and of overlap in feeding behavior (18% vs. 21%) is likewise close.

The structure of the field is the causative factor in the convergences, for as variables such as grass height and grass density change, the methods of resource division change in parallel in Chile and in North America. Such parallels exist despite the greatly divergent histories of the two areas and, as in the case of irrigated fields in Chile such as that just described, the relatively recent beginnings of some of the habitats involved. Nakamura's field (*op. cit.*) had a variety of grass heights from medium to tall, and there also the species showed partial habitat separation. *Acrocephalus bistrigiceps* occupied the tall reeds, *Emberiza furcata* and *E. yessöensis* the shorter grasses bordering water and elsewhere, and *Saxicola torquata* was found on the more barren hillsides. *E. furcata,* for example, shows a mean habitat overlap with the other species present of around 50% or a little less, just what we would expect in grasslands of the range and diversity of vegetation heights such as this.

D. Beech Forests

The north temperate forest association of beech-maple, *Fagus-Acer,* is widespread over the palaearctic and nearctic regions, and has a close parallel in south temperate *Nothofagus* forests. The southern beech is also widely distributed (Darlington, 1965), and can presently be found in New Guinea, Australia and Tasmania, New Zealand and a few smaller southwest Pacific islands, and in South America in Chile and Argentina.

The bird populations of three widely separate northern sites have been censused: Japan (Uramoto, 1961), Denmark (Joensen, 1965), and Ohio, U.S.A. (Williams, 1936). The relative abundances of the resident species are also known, and information on the feeding habits of the birds is included.

180

Similarly detailed information is available from two south temperate locations, Chile[1] (Cody, 1970) and New Zealand (Kikkawa, 1966, to which I have added pertinent data of J. Diamond, pers. comm.). Two further localities, the *Nothofagus* forests of Australia and Tasmania have also been studied by ornithologists, but only species lists are available (Kikkawa *et al.*, 1965, and Ridpath and Moreau, 1966, respectively). All census areas are alike in that the predominant tree species are beech (*Fagus* or *Nothofagus*), other broad-leaf deciduous trees are present (e.g., *Acer*), and a dense understudy of bush-type vegetation slows progress through the forest for the observer (bamboo-grass *Sasa* in Japan, the bamboo *Chusquea* in Chile and Argentina, cutting grass *Gahnia* (Cyperaceae) in Tasmania, and so on). The areas studied in north temperate sites are all similar (around 70 acres); the New Zealand site is 50 acres, but the Chilean area is only 16.5 acres.

Species can be compared among censuses for one-to-one correspondence of niches, and in the abundances of the occupants of those niches. Niches can then be grouped into guilds (see Table 14); within these groups a reduced density or absence of one species is most likely to be compensated by an increased density of another species within the group. Thus a correspondence in the densities of species within groups is expected among sites, even though the number of species in the groups does not exactly correspond. Finally the sites may be expected to correspond in numbers of species and overall population densities; the extent to which they do not may be attributed variously to a) chance effects, b) productivity differences, c) noncorrespondence of habitat, d) historical factors (man's influence, "island effects," introduced predators or competitors, etc.).

The results of the five censuses and the two species lists are summarized in Table 14. Species densities and body sizes

---

[1] Vuilleumier (1972) has produced an almost identical census from the Argentinian side of the Andes in similar *Nothofagus*.

TABLE 14. Bird species and niches in north temperate *Fagus-Acer* forests in comparison to those of south temperate *Nothofagus* forests.

Niche	Japan[1] (75 acres)	Denmark[2] (77 acres)	Ohio[3] (65 acres)	New Zealand[4] (50 acres)	Australia[5]	Tasmania[6] (16.5 acres)	Chile[7] (16.5 acres)
Sallying flycatchers low → medium → high	*Muscicapa narcissina* (76) 0.150 *Muscicapa latirostris* (72) 0.030 *Muscicapa cyanomelana* (92) 0.030	*Phoenicurus phoenicurus* (78) 0.140 *Muscicapa striata* (85) 0.029 *Muscicapa hypoleuca* (83) 0.078	*Sayornis phoebe* (87) 0.015 *Empidonax virescens* (74) 0.054 *Contopus virens* (83) 0.108 *Myiarchus crinitus* (106) 0.023	*Rhipidura fuliginosa* (74) 0.080	*Rhipidura rufifrons* *Rhipidura fuliginosa* *Petroica rosea* (part)	*Rhipidura fuliginosa* (74)	*Elaenia albiceps* (77) ½ × 0.575
	0.210	0.247	0.200	0.080			0.28
Aerial flycatchers	(*Delichon urbica*) (107)		(*Progne subis*) (142)				(*Tachycineta leucopyga*) (103)
Foliage insectivores canopy high → low	*Phylloscopus occipitalis* (62) 0.180 *Phylloscopus tenellipes* (62) 0.055 *Aegithalos caudatus* (59) 0.095	*Phylloscopus sybilatrix* (76) 0.075 *Phylloscopus collybita* (64) 0.123 *Phylloscopus trochilus* (67) 0.169 *Regulus regulus* (53) 0.007	*Vireo flavifrons* (77) 0.069 *Dendroica cerulea* (66) 0.019 *Piranga erythromela* (96) 0.119 *Dendroica virens* (64) 0.054 *Vireo olivacea* (81) 0.493	*Mohoua ochrocephala* (82) ½ × 0.188 *Zosterops lateralis* (63) 0.143	*Sericornis magnirostris* *Pardalotus punctatus* *Zosterops lateralis* *Gerygone richmondi*	*Zosterops lateralis* (60) *Pachycephala pectoralis* (100)	*Spinus barbatus* (73) 0.273 *Elaenia albiceps* (77) ½ × 0.575 *Sylviorthorhynchos desmurii* (51) 0.121
	0.330	0.374	0.754	0.237			0.682
understory	*Urosphena squameiceps* (54) 0.160 *Cettia diphone* (68) 0.165	*Hippolais icterina* (79) 0.039 *Sylvia borin* (78) 0.315 *Sylvia atricapilla* (74) 0.117	*Setophaga ruticilla* (61) 0.261 *Wilsonia citrina* (68) 0.215	*Gerygone igata* (55) 0.088 (*Xenicus longipes*)† (55)	*Pachycephala pectoralis* (*Pachycephala rufiventris*) *Sericornis frontalis*	*Pachycephala olivacea* (99) *Sericornis humilis* (62)	*Aphrastura spinicauda* (59) 0.545

	Japan	Europe	North America	New Zealand	Australia	Australia	South America
		Sylvia communis (72) 0.026					
	0.325	0.497	0.476	0.088			0.545
	0.655	0.871	1.230	0.325			1.515
Insectivores twigs & branches	*Parus ater* (59) 0.015	*Parus caeruleus* (67) 0.221	*Parus atricapillus* (66) 0.062	*Finschia novaeseelandiae* (65) 0.050	*Acanthiza pusilla*	*Acanthiza ewingii* (48)	*Anaeretes parulus* (48) 0.182
	Parus atricapillus (65) 0.120	*Parus palustris s* (62) 0.045	*Parus bicolor* (80) 0.108	*Mohoua ochrocephala* ½ × 0.188 (82)	(*Acanthiza lineata*)		
	Parus varius (76) 0.120	*Parus major* (76) 0.351			*Eopsaltria australis*		
	Parus major (68) 0.200						
	0.455	0.617	0.170	0.144			0.182
trunk surface	*Sitta europaeus* (80) 0.085	*Sitta europaeus* (86) 0.140	*Sitta carolinensis* (88) 0.054	*Acanthisitta chloris* (49) 0.280	*Climacteris leucophaea*	*Acanthiza magnus* (51)	*Pygarrhichas albogularis* (80) 0.061
	Troglodytes troglodytes (50) 0.095	*Certhia familiaris* (65) 0.159			*Climacteris erythrops*		*Troglodytes aedon* (52) 0.182
		C. brachydactyla					
		Troglodytes troglodytes (48) 0.221					
trunks high → low	*Dendrocopus kizuki* (86) 0.100	*Dendrocopus major* (141) 0.039	*Dryobates pubescens* (89) 0.046	(*Philesturnus carunculatus*)† (100)	(*Calyptorhynchus funereus*)	*Calyptorhynchus funereus* (375)	(*Dendrocopus lignarius*) (94)
	Dendrocopus major (130) 0.015		*Dryobates villosus* (120) 0.038	*Nestor meridionalis* (295) 0.010			(*Campephilus magellanicus*)† (215)
	Dendrocopus leucotis (150) 0.015		*Centurus carolinus* (131) 0.004				*Colaptes pitius* (159) 0.030
	Picus awokera (143) 0.015		*Dryocopus pileatus* (228) 0.004				
			(*Colaptes auratus*) (150) 0.004				
	0.145	0.039	0.096	0.010			0.030
	0.780	1.176	0.320	0.434			0.455

TABLE 14. (*Continued*)

Niche	Japan[1] (75 acres)	Denmark[2] (77 acres)	Ohio[3] (65 acres)	New Zealand[4] (50 acres)	Australia[5]	Tasmania[6]	Chile[7] (16.5 acres)
Ground feeders	*Phasianus soemmerringii* (218) 0.040	*Phasianus colchicus* (248) 0.013	(*Bonasa umbellus*) (175)	(*Apteryx australis*)† (188)	*Menura superba*	*Menura superba*⁺ (144)	*Turdus falklandii* (129) 0.121
	Turdus dauma (158) 0.015	*Turdus merula* (128) 0.426	*Turdus migratorius* (134) 0.015	*Gallirallus australis* (128) 0.060	*Oreocincla lunulata*	*Oreocincla lunulata* (124)	*Pteroptochos tarnii* (106) 0.121
	Turdus sibiricus (124) 0.020	*Turdus philomelus* (116) 0.218	*Hylocichla mustelina* (109) 0.300	*Turdus merula*⁺ (116) 0.055	*Colluricincla harmonica*	*Colluricincla harmonica*	*Scelorchilus rubecula* (71) 0.121
	Turdus chrysolaus (121) 0.020	*Fringilla coelebs* (90) 0.513	*Pipilo erythro- phthalmus* (89) 0.027	*Turdus philomelos*⁺ (98) 0.038	*Psophodes olivaceus*	*Turdus merula*⁺ (128)	*Scytalopus magellanicus* (49) 0.121
	Erithacus cyane (75) 0.290	*Emberiza citrinella* (88) 0.045	*Seiurus aurocapillus* (73) 0.200	*Miro australis* (90) 0.107	*Orthonyx temminckii*		
	Erithacus akahige (75) 0.005	*Erithacus rubeca* (73) 0.286	*Seiurus motacilla* (81) 0.011	*Fringilla*⁺ *coelebs* (75) 0.075	*Petroica*		
		Prunella modularis (68) 0.107		*Prunella*⁺ *modularis* (68) 0.015	*Petroica rosea* (part)		
				Petroica macrocephala (71) 0.183			
	0.390	1.608	0.553	0.533			0.363
Raptors nocturnal → diurnal	*Otus scops* (143)	*Strix aluco* (255) 0.010	*Strix varia* (333) 0.015	*Ninox novaeseelandiae* (193) 0.02	*Ninox novaeseelandiae* (200)	*Ninox novaeseelandiae* (200)	*Bubo virginianus* (340)
	Spizaëtus nipalensis (497)	(*Buteo buteo*) (383)	(*Buteo jamaicensis*) (380)	(*Falco novaeseelandiae*) (285)	(*Accipiter novaeseelandiae*) (302)	*Podargus strigoides* (260)	*Milvago chimango* (288) 0.061
		(*Accipiter nisus*) (198)			(*Aquila audax*) (598)		
Seeds & fruit		*Chloris chloris* (88) 0.065	*Richmondena cardinalis* (94) 0.073	*Cyanorhamphus auriceps* (119) 0.160	*Aprosmictus scapularis*	*Platycercus caledonicus* (183)	*Microsittace ferruginea* (179) 0.061
		Carduelis cannabina (80) 0.013		*Carduelis flammea* ( ) 0.075	*Platycercus elegans*		

Omnivores	Cuculus saturatus* (200)	Coccothraustes coccothraustes (103) 0.016	Hedymeles ludovicianus (101) 0.008	Chalcites lucidus* (103) 0.040	(Cacomantis pyrrhophanus)*	Kakatoë galeria (370)	Columba araucana (210) 0.061
	Sphenurus sieboldii (185) 0.010	Columba palumbus (245) 0.169	(Coccyzus americanus)* (144)	Eudynamis taitensis (192) 0.027	Chalcites basilis)*		Curaeus curaeus (128) 0.303
		Columba oenas (223) 0.049	Ectopistes migratorius† (205)	(Turnagra capensis)† (131)	Acanthochaera carunculata	Phaps elegans (160)	
	Garrulus glandarius (170) 0.065	Oriolus oriolus (156) 0.055		Hemiphaga novaeseelandiae (259) 0.02	(Leucosarcia melanoleuca)	Strepera fuliginosa (255)	
		Garrulus glandarius (179) 0.003	(Cyanositta cristatus) (132)	Prosthemadera novaeseelandiae (147) 0.075	Strepera graculina		
	0.085	0.370	0.081	0.322			0.425
Nectivores				Anthornis melanura (88) 0.163	(Acanthorhync tenuirostris) (Meliphaga chrysops)		Sephanoides sephanoides (64) 0.121
Scavengers		Corvus corone (332) 0.013	(Cathartes aura) (535)				Coragyps atratus (538) 0.091

* Introduced
* Brood parasite
( ) Not in census, but known to be present in area
¹ Data from Uramoto (1961)
² Joensen (1965)
³ Williams (1936)
⁴ Kikkawa (1966) and J. Diamond, pers. comm.

† Rare; nearly extinct
†† Extinct

⁵ Kikkawa et al. (1965)
⁶ Ridpath and Moreau (1966
⁷ Cody (1970)

(total length in mm) are also included. Each species is aligned as closely as possible with its counterpart(s) in other forest habitats, although in a good number of cases this is not easy (nor even possible). Some "split appointments" are made for more accurate niche representations. The niche associations are listed from those I consider most distinct—the sallying flycatchers—to those in which associations are made only tentatively and at considerable risk of error. The last major category, "Seeds & Fruit, Omnivores," is to some extent a catchall grouping, comprising as it does the true seed and fruit eaters of the canopy, the omnivorous jays and their homologues, the insectivorous cuckoos, the pigeons and doves, and finally the oriole-equivalents. Elsewhere in the table, nicheassociations are shown in an intended sequence, with flycatchers and foliage (canopy and understory) insectivores adjacent and likewise the birds which hunt over twigs, branches, and trunk surfaces.

There are obvious qualitative similarities among the species lists, and likewise there are some anomalies which might at first sight shake our faith in natural selection. To what extent do fewer species per niche grouping compensate by increased density per species for the absence of competitors? For each of the ten pairs of habitats, I obtained the correlation coefficients r_{sp-sp} of the densities of species in one census with the densities of their putative replacements, the species which provide the best possible one-to-one matches, in the second locality. These correlations coefficients averaged 0.356 (range 0.10–0.58, 4/10 significant). A second set of correlation coefficients r_{gr-gr} was obtained by comparing densities between niche groupings rather than species to species. Higher values were scored (mean = 0.512); range 0.2.–0.83 3/10 significant). The ratio r_{gr-gr}/r_{sp-sp} is a measure of the degree to which species compensate within niche groupings for more or fewer species, and is independent of productivity considerations; the ratio exceeds unity in eight out of ten pair-wise comparisons, in some cases by a great margin. The extent to which this compensation takes place is in part due to the discrepancies in the numbers of species available to fill the niche space. The more similar are species counts per locality, the less opportunity to exercise density compensations by the resident species. Accordingly, when the ratio of higher to lower species counts per census is plotted against the ratio of grouped to single niche density correlation coefficients, a positive relation is obtained (Figure 54).

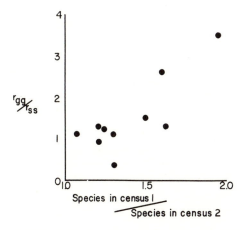

FIGURE 54. In beech forests the densities of ecological analogues are correlated between species (r_{ss}) and between groups of ecologically similar species or guilds (r_{gg}). The extent to which the between-census correlation is improved by grouping species into guilds (r_{gg}/r_{ss}) is proportional to the extent that one census is species-depauperate with respect to the other.

The beech forest data, while crude, serve to illustrate various factors that influence species density. Historical or chance effects must be mainly responsible for the differences in species numbers within the niche groupings and in the census totals. The total density of pairs/unit area varies from a low in New

Zealand of 1.857 pr/ac to a high in Denmark of 4.485 pr/ac and appears to be largely independent of species numbers. How important is this overall density figure in the absolute densities of niche groupings? The sallying flycatchers appear remarkably constant in density among four habitats (the exception is a low count from New Zealand), indicating, perhaps, that an upper density limit has been reached at each site, regardless of the number of species (one to four) which compose this guild. Other niche associations, such as the canopy and understory insectivores, are variable in density from site to site; in former but not in the latter, density increases with

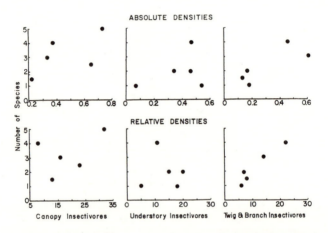

FIGURE 55. Comparison between five intercontinental beech forest censuses. Three niche groupings, or guilds, are considered: canopy insectivores, understory insectivores, and twig-and-branch insectivores. Number of species/guild may or may not be correlated with absolute and relative guild densities.

the number of species per niche association (Figure 55, a and c). When a correction is made for overall species density by expressing group as a percentage of the total plot density, the scatter remains for the understory species and increases for the canopy groups (Figure 55, b and d). Contrary to this, the insectivores of twigs and branches show a somewhat orderly relation between numbers of species and absolute guild density, which when expressed in terms of relative guild density resolves into a positive linear relation (Figure 55, e and f). Thus the density of the guild D_{gr} is directly proportional to both the number of occupant species n and the overall importance P of the guild in the community: $D_{gr} = 0.05nP$ in this case. Little of more sophistication is worth attempting with these data; suffice it to say that the accidents of history and chance may be partially compensated by niche expansion, which is reflected by increased densities per species when fewer species divide resources, and may or may not be influenced by relative guild importance, expressed as the percentage of community bird density contributed by the guild.

E. Mediterranean Scrub

1. The Habitat. The chaparral association of California, composed predominantly of fire-adapted broad-leafed sclerophylls around 2–3 m in height, parallels in many aspects of physiognomy, morphology, flowering seasons, and other adaptive features the matorral of central Chile (Mooney and Dunn, 1970; Di Castri and Mooney, 1974).

In 1968, I selected a Chilean matorral study area on the basis of its physiognomic similarity to the chaparral site in the Santa Monica Mountains. The site chosen was 4.6 acres at 900′ elevation on the coastal range near Puchuncavi, 75 km north of Santiago. The dominant plant species there are of *Lithraea* and *Quillaja,* and their convergent similarities with Californian shrubs have been documented at some length (Di Castri and Mooney, 1974). In both plots the vegetation is virtually continuous, a close-canopy shrub cover of variable height but of constant high density. Vegetation height distributions for the sites are given in Figure 56. The Chilean plot was censused throughout the 1968 breeding season, and com-

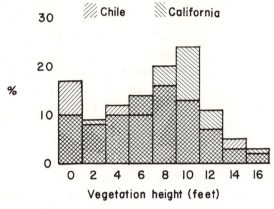

FIGURE 56. Frequency distributions of 50′ × 50′ quadrats by vegetation height, in Californian chaparral and matched Chilean matorral.

parable information on its use by bird species to that of the California site was obtained. This information includes: a) numbers and kinds of species, b) their distribution and abundance within the study areas, c) their feeding height distributions within the vegetation, and d) quantitative measurements of their feeding behaviors.

The California chaparral is intermediate on a moisture gradient between the drier coastal sage association (*Artemesia-Salvia-Eriogonum*) and the moist riparian or oak woodlands (*Quercus agrifolia, Umbellularia, Juglans*). In Chile a similar ordering of habitat occurs, with matorral intermediate on a habitat (moisture) gradient between coastal *Bahia-Happlopappus* and inland *Cryptocarya* woodlands. Central Chile has another habitat, "savanna," with no counterpart in California. This is a low open-savannah-like habitat chiefly of *Acacia-Prosopis,* and may have been derived from matorrallike vegetation by the influence of man and his goats.

Table 15 gives the census results and shows guilds of ecological counterparts based on habitat and feeding ecology characteristics.

2. Species Distributions over the Habitat Gradient. Table 15 lists only those species which may be seen in the primary habitat, chaparral or matorral, in the breeding season; names are spaced horizontally according to where each species reaches its maximal abundance in the habitat gradient.

Very few species are restricted to the primary habitat type. The wrentit *Chamaea fasciata* and the California thrasher *Toxostoma redivivum* are common only in chaparral, and *Asthenes humicola* (the wrentit homologue) plus perhaps *Phytotoma* are common only in matorral. About the same number of the species listed in each locale reach maximal abundance to the woodland side of center in the gradient, but the matorral contains more birds relative to chaparral which are commoner in drier habitats. To some extent this is a reflection of the greater proportion of open ground (17% vs. 10%) and of lower mean height of vegetation (6.3 ft vs. 8.7 ft) in the matorral as compared to the chaparral site. On average, the Chilean species are more widely distributed over the habitat gradient (+10–20%). Thus there are 22 unlisted species which occur in oaks or coastal sage but not in chaparral, and only 8–10 such Chilean species, in spite of the fact that the number of species per habitat is at least as great in Chile.

3. Status in the Censuses: Full Members and Marginal Species. Only those species whose distributions are actually intersected by the dashed line in the table were resident within the census areas ("full members"). Some full members of one census have obvious counterparts in the other locale which were recorded as "marginal" there (part of a territory within census area; breeds adjacent to census area). Such marginal species are also listed in the table. Thus in chaparral the western flycatcher *Empidonax difficilis,* common in oak woodlands,

Table 15. Census results and species correspondence between two Mediterranean scrub sites.

	California			Chile		
#	Coastal sage	Chaparral (6.9ac)	Oak woodland	Woodland	Matorral (4.6ac)	Bahia-Happlopappus
1	118/0.71[1]	Thryomanes bewickii →		←—— Troglodytes aedon ——→		120/0.65
2	151/1.03	Chamaea fasciata		←— Asthenes humicola		165 0.33
3	102/0.65	←—— Psal|triparus minimus ———→		←— Leptasthenura ae|githaloides		160/0.22
3a	108	|←Polioptila caerulea—→		Aphrastura spinicauda ——→		140
4	127/0.16	Parus inornatus—→		←— Anaeretes parulus—→		110/1.09
5	146/0.03	|←—Empidonax difficilis ——→		←—— Elaenia albiceps ——→		150/0.33
6	197/0.13	←——Myiarchus cinerascens——→		←— Pyrope pyrope ——→		210/0.27
7	96/0.45	←—Calypte|anna ——→		←—Patagona|gigas—→		212/0.45
7a	86	←—Archilochus alexandri——→		←——Sephanoides sephanoides|		110
8	250/0.39	←——— Lophorty|x californica ———→		Notho|procta perdicaria——→		305/0.04
9	205/0.32	←—Pipilo fuscus——→		←———Lophortyx californica——→		250/0.22
10	138	Carpodacus mexic. |		—Diuca diuca——→		175/0.33
				—Zonotrichia capensis——→		147/0.82

No.	Size/density[1]	California		Chile	Size/density[1]
11	540	Geococcyx ——→	—————Pteroptochos megapodius		232/0.22
12	298	Zenaidura macr. —→|	·———————Zenaidura auriculata		265/0.33
13	289/0.45	↓—Aphelocoma coerulescens—————→	↓—————Agriornis livida——————→		275/0.11
14	199/0.52	↓—Pipilo erythrophthalmus—————→	—Turdus falklandii———————→		280/0.33
15	284/0.45	Toxostoma redivivum———————→	↓—Mimus tenca———————→		290/0.44
16	181/0.19	Pheuticus| melanocephalus————→	↓—Phytotoma| rara——→		195/0.33
16a	103	↓——————Spinus lawrencei———→	↓—Spinus barbatus————→|		135
17	184	↓———Dendrocopus nuttallii———→	↓—Dendrocopus lignarius———→|		182
18	280/0.10	↓——Colaptes cafer——————→	↓—Colaptes pitius————→		320/0.02
19	147	↓————Aeronautes saxatalis————→	↓————Tachycineta leucopygia—————→		135/0.44
20	197	Phalaenoptilus nuttallii————→	↓————Chordeilles longipennis—————→		243
		↓—Bubo virginianus———→	↓——Bubo virginianus——→		
		↓—Cathartes aura———→	↓—Cathartes| aura——→		
		↓———Buteo jamaicensis———→	↓—Parabuteo unicinctus——→		
			Accipiter cooperii———→	↓—Milvago chimango——→		

[1] Body size/density.

occurs on the edge of the study plot and occasionally feeds within it. Its Chilean counterpart, *Elaenia albiceps,* is a full member of the matorral census. The reverse is true for the woodpeckers *Dendrocopus.*

4. Ecological Counterparts. At once it can be seen that the two bird communities parallel each other with remarkable precision. Almost every full member species in each census can be matched with one-to-one correspondence to species in the parallel habitat, and 20 pairs of suggested ecological counterparts are numbered (Table 15). Of these, 15 involve both full member species, and five associate a full member species (four Chilean, one Californian) with a marginal species. In each census, four foliage insectivores, two flycatchers, two woodpeckers, one nectivore, one seed-fruit eater, and both diurnal and crepuscular aerial feeders occur. The one-to-one matching inspires less confidence with the ground-feeders, where some pairings are more anomalous than homologous. Only one full member species is not matched up, the Chilean tinamou *Nothoprocta perdicaria.* It is displaced in the matching process by the quail *Lophortyx california,* which was introduced into Chile from California in 1870. Three numbered pairs are of matched marginal species, and each site also supports a large owl, a scavenger, and both searching and pursuing hawks.

Some of the species pairings provide near-perfect matches, the more interesting of which are developed below. Others, especially among ground feeders, are not so perfect. When the taxonomic differences between censuses are considered, the two communities duplicate each other with impressive fidelity. The 23 numbered matches include, after all, nine between the members of different families and eight between confamilials, with only five between congeners plus one conspecific, the Chile-California quail. If the paired species in Table 15 are ecological counterparts, then we would expect their morphologies and their general ecologies to coincide, almost by definition. This can be tested in a simple way by plotting Chilean and California values of various indices against each other. This is done in Figure 57 for body size. The correlation

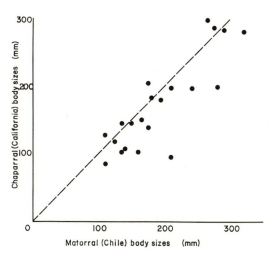

FIGURE 57. Correspondence between chaparral and matorral ecological counterparts in body size.

is good ($r = 0.66$), and the slope of the regression line is very close to unity ($b = 1.05$). An interesting across-the-board increase in body size in Chilean over California species shows up ($a = 12.6$ mm), which can probably be attributed to the differences in lineage between the occupants of the two sites, and which has not quite been erased by natural selection. Surprisingly, the correlation coefficient for a similar plot of bill lengths rather than body sizes indicates a slightly poorer correspondence ($r = 0.61$).

General ecological requirements are reflected in average territory sizes or in densities (pr/ac). The matched species show a considerable scatter in one-to-one density correspondence (Figure 58). Part of this scatter is resolved when niches are grouped and densities summed, but the Chilean site maintains a greater productivity with 6.5 pr/ac vs. 5.6 pr/ac for the

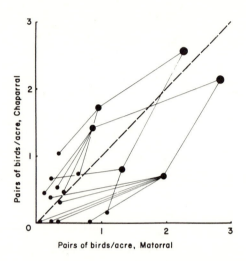

FIGURE 58. Correspondence between chaparral and matorral ecological counterparts, by species-species (small dots), guild-guild (medium sized dots) and census subtotals (large dots).

chaparral (17 measured niche densities). Mean feeding heights between matched pairs are very similar ($n = 12$; $r = 0.85$), in spite of a poor match between the height of their selected habitats ($r = 0.17$).

The two 20×20 community matrices of the Californian and Chilean bird communities containing all pair-wise ecological overlaps are given in Appendix B, and these are then used to construct the dendrograms of ecological affinities in Figures 59 and 60.

Inspection of the matrices shows that overlaps generally increase horizontally and vertically away from the diagonals,

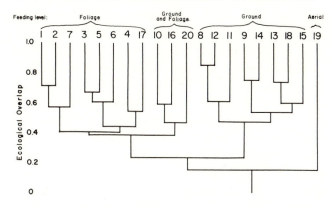

FIGURE 59. Dendrogram of ecological relations among 20 chaparral bird species. Species numbers as in Table 15.

supporting the original grouping in Table 15. Species are grouped into similar arrangements in the two community dendrograms, with two major associations of foliage-level and ground-level feeders. An extra group is formed in chaparral by the four species fairly restricted to the tallest vegetation (live oaks), but the rank correlation of the species numbers

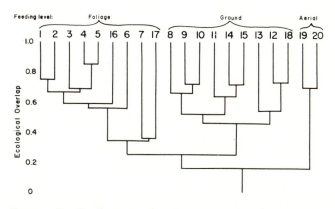

FIGURE 60. Dendrogram of ecological relations among 20 matorral bird species. Species numbers as in Table 15.

across the top of the two dendrograms remains very high ($\Sigma d_i^2 = 106$, $n = 17$, $p < 0.01$).

5. *The Foliage Insectivores and Sallying Flycatchers.* The canopy insectivores may be treated in further detail. These species, grouped into the first two categories of Table 15, number six at each site and correspond well in total density (2.7 vs. 2.9. All are full members except *Empidonax difficilis,* which is marginal at the chaparral site. The four foliage insectivores are mainly separable on the basis of foraging height. In chaparral, the wrentit *Chamaea fasciata* feeds low in the vegetation (average 3.09') and to some extent on the ground. Above it may be ranked Bewick wren *Thryothorus bewickii* (average 3.30'), bushtit *Psaltriparus minimus* (average 4.47'), and plain titmouse *Parus inornatus* (average 8.0'). In Chile the homologues are similarly stacked in feeding zones, with *Asthenes* in the lowest position (average 1.71'), *Troglodytes* above it (average 3.88'), and *Leptasthenura* and *Anaeretes* in the highest positions. The latter two occupy reversed position with respect to the chaparral species (averages 7.76' and 4.20' respectively), and show the effect of decreased vegetation height at the matorral site. In each plot the larger flycatcher feeds at higher levels than does its smaller competitor and, in each country, is characteristic of drier and more open vegetation while the smaller flycatcher is more abundant in forests. Figure 61 gives the feeding height histograms using six height categories; chaparral species have a reduced spread of activity among these categories as compared to the matorral species, for a diversity index of their coverage of vegetation height averages 2.55, only 78% of the 3.25 figure for their Chilean counterparts.

Silhouettes of the 12 species are included in Figure 61 and are drawn to scale; the phenotypic correspondence in size and proportions is very striking. Lastly, the feeding behaviors of the species are compared in Figure 62. The matorral species

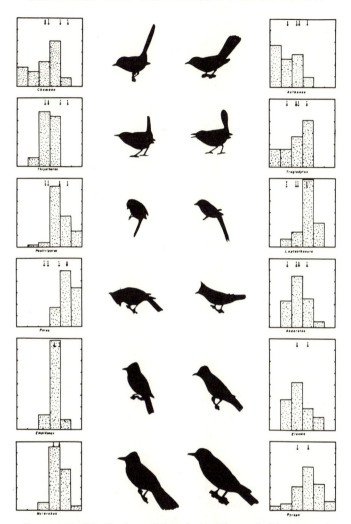

FIGURE 61. Distributions of foraging heights and silhouettes of four canopy insectivores (upper) and two sallying flycatchers (lower) in Californian chaparral (left) and Chilean matorral (right). Foraging height intervals are, from left, ground, ground–6″, 6″–2′, 2′–4′, 4′–10′, and 10′–20′. Silhouettes are all to the same scale. Arrows above each distribution indicate its mean and those of the frequency distributions of neighboring species.

199

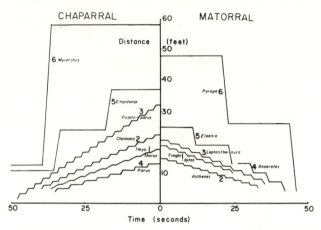

FIGURE 62. Feeding behaviors of the six pairs of ecological counterparts shown in Figure 61 from Californian chaparral and Chilean matorral.

are rather less diversified than the chaparral counterparts, which is perhaps associated with their broader foraging height niches. The flycatchers provide distinctive behavior graphs and correspond well, but the foliage insectivores are not as easily matched. There are evidently, at this level of detail, a variety of efficient ways, bracketed by the titmouse and bushtit feeding behaviors, in which birds can move through scrub vegetation and harvest insects.

In summary, the average matorral species differs from the average chaparral species in a) broader occupancy of the "coastal-sage"-woodland habitat gradient (Table 15), b) broader range of vertical feeding zones (Figure 61); the average matorral species pair differs from the average chaparral species pair in a) 3% more habitat overlap (75% vs. 72%), and b) 4% more overlap in food/feeding behavior (52% vs. 48%), but foraging height overlaps are the same. These differences may be regarded as minor in comparison with differences between communities in general (see Table 5).

7. *Chaparral Elsewhere.* It would be extremely interesting to extend the Chile-California comparison to the bird communities of the same vegetation type elsewhere. Unfortunately almost nothing has been published on the southwestern Australian equivalent, and only bird species lists from South Africa are available (Winterbottom, 1966; Broekhuysen, 1966). Some comprehensive studies have been conducted in *garrigue* in southern France by Blondel (1965, 1969), but his site appears to support a rather lower and more patchy vegetation, and is extremely low in bird species. Further work with careful habitat matching should prove to be very valuable.

IV. CONVERGENCE BETWEEN AVIFAUNAS

The final question to be considered in this chapter is: Does the whole avifauna of a particular region show any correspondence to that of other, comparable regions? The answers to this question are obscure for several reasons. If the question is to be at all meaningful, the way in which "regions" should be "comparable" is in habitat types and perhaps also in the relative proportions of each habitat type. Such matches between large geographic areas are correspondingly difficult to make, and the conclusions drawn are less valuable.

I have attempted (Cody, 1970) this sort of comparison between the nonmarine bird faunas of California and Chile, although in a superficial way. These are two of the very few regions worth comparing perhaps, for California and Chile are similar in size and latitudinal range, both face an ocean and are backed by tall mountains, and both possess a range of habitats from dry deserts at low latitudes through Mediterranean habitats to wet forests at higher latitudes. Each region also possesses a coastal mountain range and prominent north-south valleys.

The total number of species in each region is extremely similar, with 235 in California and 230 in Chile. However, the distribution of these species among habitats within each region differs, as the Californian species tend to be habitat-specific and occur over the whole state (inasmuch as do their specific habitats). In Chile, on the other hand, species are much less habitat specific but are more restricted in geographic distribution; a species turnover occurs with shifts on a north-south axis rather than between habitats.

A different approach to avifaunal equivalence was taken by Lein (1972), who examined the occupancy in different zoogeographic areas (Sclater regions) of various trophic levels in terms of numbers of species. Correspondence was high (90%) between regions of similar habitat types (e.g. nearctic and palaearctic) and of close proximity and historical association (e.g. nearctic and neotropics), and reduced to around 70% to other pair-wise comparisons. Several anomalies showed up, such as a low incidence of nectivory in the palaearctic, and a scarcity of ground-foraging birds in Australia, but the overall similarities in the numbers of species which exploit different sets of food resources remain impressive.

Less comprehensive comparisons, for instance between homologous families, have been drawn, and some have already been mentioned (e.g. Fry, 1970, and those in Section II.C.1). Lack (1968, Ch. 5) has shown that many parallel species can be identified in the Icteridae (North and Central America) and the Ploceidae (Africa), and discusses many more instances of convergent evolution at lower taxonomic levels. These comparisons, and indeed most of those made between families and avifaunas are necessarily of a qualitative nature, and lie more in the realm of comparative and historical zoogeography than with community ecology. Pending more quantitative and detailed studies of these higher groups, I consider them to be of reduced value in furthering the conclusions already made.

CHAPTER SIX

Alternatives To Competitive Displacement Patterns

I. LIMITATIONS TO DISPLACEMENT PATTERNS

In this final chapter, alternatives to the quasi-stable displacement patterns discussed above will be examined. First, factors which tend to reduce the prominence or conspicuousness of these patterns are mentioned; and second seemingly nonequilibrial displacement patterns will be described in which pairs of species are interspecifically territorial.

A. *Nonlimiting Resources*

Displacement patterns depend for their existence on selection against individuals which depart from the species mean in the utilization of a resource. This departure brings such individuals into competition with members of adjacent species on the resource span and is only a disadvantageous trait if the resources involved are in short supply. Thus, no displacement patterns can evolve on superabundant resources, which can be simultaneously and similarly used by several species.

There are instances in bird communities where resources are apparently nonlimiting and are used by many species in similar ways. Large numbers of individuals of several species congregate around temporarily abundant food supplies such as fish schools, the insect prey fleeing grass fires (Chapin, 1932, over a dozen species in nine families), on flowering and fruiting trees in the tropics (Land, 1963; Willis, 1966; Diamond and Terborgh, 1967; Terborgh and Diamond, 1970), and around garbage dumps on land and sea. The number of seabirds of many species which gather round fishing boats when the catch is cleaned is an example of the last; near Iceland, these may include at least four gulls (*Larus*), kittiwakes (*Rissa tridactyla*), jaegers (*Stercorarius spp.*), and fulmars (*Fulmaris glacialis*). Local and sporadic areas of upwelling inshore waters also attract large numbers of individuals of many species which would not ordinarily be found feeding adjacent to each other. Such upwellings off Destruction Island, Washington State, provide food for two wintering southern hemisphere shearwaters (*Puffinus*), five local species of Alcidae, a gull *Larus glaucescens*, a jaeger *Stercorarius*, and two species of cormorants, *Phalacrocorax pelagicus* and *P. auritus*. The common features of a) the short duration of the food sources, b) its temporary abundance, and c) its relative unpredictability in space and time apparently rule out the possibility of the evolution of displacement patterns among the exploiting species. These factors which prevent *interspecific* resource allocation are evidently the same as those which prevent *intraspecific* resource division by means of territoriality in the customary way. Rather, they favor colonial or social food exploitation by nonrelated individuals within the species (Horn, 1968).

Two results of these types of food supply, intraspecific sociality and a lack of interspecific ecological divergence, may be observed simultaneously in African vultures. Six species are found in the Serengeti (Kruuk, 1967). Their food supply occurs in large servings (whole or parts of game carcasses) of short duration (due to the return of the predator which killed it or to decomposition), and which occur sporadically, sparsely, and unpredictably over large areas. Most species are gregarious if not strictly social, and seem to have no intraspecific spacing mechanisms which apportion hunting areas. All six species may gather at a single carcass, and morphologically, behaviorally, and in terms of the parts of the carcass they prefer to eat, may be broken down into three pairs (Figure 63).

Aggression at the carcass is most common within species, but aggression between the members of the pairs of ecologically related species is also frequently observed. A displacement pattern in size, ability to rip hides, tear offal, etc., exists among the pairs of species, and is correlated with the order in which the species arrive at the kill.

Within these three species pairs, little can be discerned which might promote their coexistence. The species not only fail to show both the usual avian intraspecific and less usual interspecific spacing mechanisms, but rather they appear to have evolved ways of coordinating their common and similar use of the carcass. The dipping flight, which signals to all vultures within sight that a bird has spotted food, and the wing patterning that species of similar ranges tend to hold in common ensure that a find attracts maximal attention from all individuals of all species from many miles around. The system makes sense if some sort of succession in the use of the carcass takes place between various species, i.e., one rips the hide and eats subcutaneous tissue, thereby exposing the viscera and soft flesh for a second species whose ability to handle that food is superior but whose hide-ripping talents are limited, and which eventually gives way to the bone-pickers; but still the coexistence of the species within the three pairs warrants explanation. Interestingly enough, these paired species are those with largely complementary geographical ranges. In their limited areas of sympatry, it seems that selection favors ecological convergence to bring species pairs to act as a single ecological entity, which employs such individual spacing mechanisms as aggression at the carcass just as would a single species.

Data exist on seabird associations which also bear on the relation between abundance of food supply and the evolution of displacement patterns among its users. Belopolsk'ii (1957) reported that, in years of normal fish abundance, fish were the major diet item in five Alcidae, in gulls and cormorants in the East Murman. However, during the occasional years

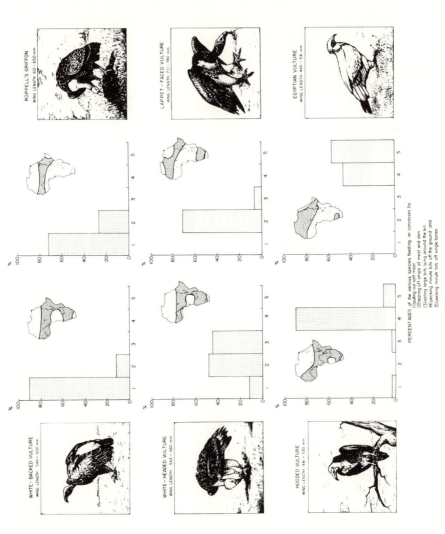

RÜPPELL'S GRIFFON
WING LENGTH 612-650 mm

LAPPET-FACED VULTURE
WING LENGTH 710-780 mm

EGYPTIAN VULTURE
WING LENGTH 445-531 mm

WHITE-BACKED VULTURE
WING LENGTH 544-600 mm

WHITE-HEADED VULTURE
WING LENGTH 563-660 mm

HOODED VULTURE
WING LENGTH 416-550 mm

PERCENT AGES of the various species feeding on carcasses by
(1) pulling out soft meat
(2) tearing off strips of meat and skin
(3) eating large bits lying around the kill
(4) pecking minute bits off the ground and
(5) pecking minute bits off large bones

FIGURE 63. See legend on opposite page.

of low fish abundance some species were "squeezed out," leaving a series of more specialized fish-eaters that differ from each other chiefly in the distance out to sea at which they forage; the excluded birds were forced to eat crustaceans. Thus high resource abundance allows more species to exploit a resource, and low abundance precipitates the exclusion of more poorly adapted species from that resource onto other, presumably less profitable, resources. Ingolfsson (1970), in a study of large Icelandic gulls, was able to show that gull species' diets were quite different in foods which occurred regularly and predictably throughout the season, but did not differ in food items which were sporadic in occurrence. In Alaska, lemmings support many different predator species only in years when they are at peak abundances (Pitelka *et al.*, 1955), and similarly the only diet item shared among German raptors is the vole *Microtus,* which is also the only prey species to become periodically superabundant there (Lack, 1946). These results adequately summarize the relation between the predictability of a food supply and the orderliness with which it is subdivided among the species it helps to support.

B. Nest Sites

Displacement patterns are apparently discernible on resources which should logically be unlimited, especially when compared to other resources which are far more likely to be exhausted first. Nest sites are an example, although these may be limiting in species with special requirements, such as hold nesters (see Lack, 1966, and the work of his students) or the open country raptors (mostly *Buteo* hawks) which use virtually every one

---

FIGURE 63. The feeding habits and ranges of six species of vultures in eastern Africa. The six species break down into three pairs, with ecological differences between the pairs but not within them. Members of a pair are similar in size and feeding habitats, but differ in geographic range characteristics and in wing patterning. Data adapted from Kruuk, 1967.

of the cottonwood trees which dot the short-grass plains in northeastern Colorado. But there should be an abundant choice of sites for canopy, ground, and seacliff nesters. The question has received much comment with regard to seabirds, where, unfortunately, confusion is added because of the potential value in population self-regulation of strongly traditional colony sites. Leaving such considerations aside, many students of seabirds have commented on interspecies segregation of nest sites on the same cliff or coastline (e.g., Naumann, 1903; Lack, 1934; Sergeant, 1951; Nørrevang, 1960; Bedard, 1969a, 1969b). Bedard (1969a) studied three alcid species that nest on St. Lawrence Island in Alaska. These species, parakeet auklet *Cyclorrhynchus psittacula,* crested auklet *Aethia cristellata,* and least auklet *Aethia pusilla,* display a body size series (7¼″, 7″, and 5¼″, respectively) which, together with differences in bill proportions, results in or at least coincides with corresponding differences in prey sizes (all three species feed on macroplankton). Nest sites are selected among the boulders of rocky slopes according to distinct species-specific preferences for substrate rock sizes. It is quite likely that this pattern of nest site segregation has come about not through direct interspecific competition for different sizes of crevices, but rather that it is simply a by-product of different body sizes which have been selected relative to feeding ecology. Of course, a burrow which is a close fit is optimal, as larger burrows increase the risk of predation.

The species-specific differences in nest site among the six North Atlantic alcids have often elaborated, and are supposed to be the results of interspecific competition (e.g., Lack, 1934). Indeed, some observers record aggression between species which might well serve to maintain the displacement pattern (Bedard, 1969c; Belopolsk'ii, 1957); the larger species usually dominates the smaller and secures the nest site.

It may be that such behavior and its results are more profitably viewed from a rather different angle, that of the ultimate factors which determine optimal site selection in these species. I have commented (Cody, 1973) on nest site selection in the six-species alcid communities found on the coasts of Grimsey, northern Iceland, and on the Olympic Peninsula, Washington State. Site selection is interpreted, along with other factors in feeding and breeding biology, in terms of two sets of selective forces: a) interspecific competition for feeding areas at sea, and b) predation, on both adult and chick. The former selection pressure produces a zonation of feeding areas with respect to distance from the breeding rock. Because of depletion of food in the heavily fished inshore areas and the advantages of having the chick close to the fishing parent, the young may profitably leave the nest sites at weights as low as 20% of the adult. This behavior necessitates an open, exposed nest site with good sea access, but which is therefore exposed to predatory gulls. The attentions of these marauders are discouraged by the presence of one of the parents, which in turn is economically permissible by the proximity of the food supply; parental guard duties are in addition feasible because the same proximity of food supply selects for large body size in the inshore-fishing species, adequate for successful deterrence of gull predation. Further, inshore-fishing species must be least selective in nest sites if all the available inshore waters are to be utilized with no costly increase in the distances the parents fly with food for the young.

Species which feed offshore, on the other hand, cannot afford to expend one parent in guard duties, as their chicks, even with both parents fishing for them, grow exceedingly slowly. Thus the chicks must be sheltered in deep crevices or in burrows; colony site selection is therefore more discriminating both because of the more rigorous nest site requirements and because wider dispersal of colonies (of a larger size, accordingly) adds proportionately little to the total foraging distance of parents fishing many miles offshore. It seems, then, that the behavioral reinforcement of nest site selection described above may ultimately have its selective basis in feeding zonation and chick protection.

C. Length of Growing Season

For twenty or so years after Gause (1934), it was a practice among ecologists to identify species assemblages that persist in apparent contradiction to the Volterra-Gause principle. Eventually the species in most such assemblages were shown to differ after all, in some subtle and unsuspected way or in some novel niche dimension. For instance, the "problem" of coexistence of the New England wood warblers, posed by Kendeigh (1947), was resolved by MacArthur in 1958. But Hutchinson (1941, 1961) raised and focused a problem—"the paradox of the plankton"—which has not been disposed of by subsequent theorizing or experiment. He asked: How is it that so many phytoplankters are able to coexist in such an apparently unstructured, homogeneous environment as the surface waters of lakes and oceans? The competitive exclusion principle leads us to expect, given that the description of the environment is accurate, that one species would prove competitively superior and would eventually monopolize the habitat. By allowing some habitat structure, through perhaps light and chemical gradients, at most a few species should persist. Arguing that quite limited niche diversification is possible through commensalism and mutualism, and by minimizing the role in this coexistence of limitation of predators, Hutchinson concluded that one of the basic assumptions of competitive exclusion logic is not warranted—that the plankton communities do not, in fact, reach equilibrium. Further, such nonequilibrium situations are to be expected whenever the time scale for environmental changes, which may reverse the direction of the competitive process, is neither much less than nor much more than the generation time of the organisms in competition, but is approximately of the same order of magnitude.

In plankton communities, the common species are more

common and the rare species are rarer (the type "IV" distribution) than expected from the MacArthur (1960) broken stick model. As this model assumes that a competitive equilibrium has been reached, the observed abundances may be further evidence that such a state has not been reached in the plankton. These type IV distributions of relative abundances might well arise if the environment temporarily favors one organism over the others for a short while but soon switches to promote another—a temporal analog of spatial heterogeneity.

Williams (1971) presents another alternative. Different species might stagger their nutrient uptake by growing at different rates, perhaps on a daily cycle. Even though the same net growth rate were attained in all species, the differences in the times of peak resource requirements among species might promote coexistence among species with different cell cycle patterns (species might grow linearly, exponentially, or in some other power function fashion to achieve equal net growth/day).Yet, once several species with hypothetically identical nutrient requirements are growing together simultaneously, overlaps in the timing of nutrient uptake rapidly become great. Further selection will favor, in the species which reaches peak nutrient uptake soonest, a modification of its late-cycle growth stages to compete more effectively with and eventually exclude other "late-blooming" species. Something has been gained, but the phytoplankters remain paradoxical.

The relation between the length of time one set of environmental conditions remains unaltered and the extent to which competitive displacement occurs can be seen more clearly in bird communities, although there too the results are far from transparent. Figure 64 contrasts the niche overlap in habitat, in feeding heights and in feeding behavior between paired migrant and paired resident species at the North American sites. Migrants show the same or slightly less habitat overlap α_H among themselves compared to resident species, regardless

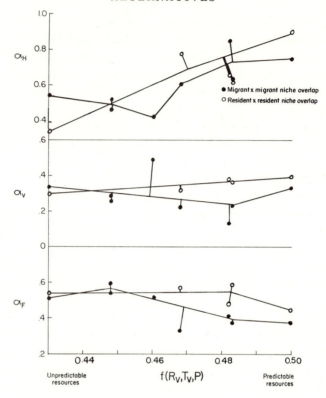

FIGURE 64. Contrast between variation in niche overlap components with climatic predictability between migrant species and between resident species.

of the predictability of the habitat's resources. In unpredictable habitats, α_V and α_F are likewise similar between residents and migrant species pairs, but in predictable habitats residents appear to tolerate greater niche overlaps in these components. These data indicate that residents with prolonged contact with each other evolve to higher niche overlaps than those between species with shorter contacts. In general in habitats with long breeding seasons habitat overlaps are greater and food/feeding behavior overlaps are reduced ($r_{FFD,\alpha_H} = 0.75$; $r_{FFD,\alpha_F} =$

-0.75); foraging height overlaps and breeding season length are not related ($r_{FFD,av} = -0.18$).

D. Limitation by Predators

If populations of ecologically related species are limited by predators as, for instance, herbivores might by (Hairston *et al.*, 1960), then in the absence of food limitation they may not exhibit displacement patterns of food resources. Generalities at this broad level are perhaps not worthwhile, but there is little if any evidence that the sorts of space and food segregation mentioned earlier in this work are more restricted to secondary consumers than to primary consumers. The division of seeds of various types among ploceid finches (White, 1951), and the interrelations among the herbivorous ungulates of African savannahs (Lamprey, 1963; Vesey-Fitzgerald, 1965), indicate that competitive displacement applies equally to trophic groups below insectivores and carnivores.

Yet predators have been shown to influence coexistence among their prey species. In some cases predation apparently keeps prey population levels below those at which competitive exclusion occurs. Darwin (1859, Ch. III) mentioned the case of lawn plants, observing that, if lawns are kept mowed, many species persist in them. But if the lawn is neglected (not cropped or "predated") fewer species can survive—an initial 20 species was reduced to 11 after cessation of mowing. Likewise in sheep-grazed pastures grasses of both *Festuca* and *Poa* may coexist, but, when sheep are removed, the former comes to predominate. Paine (1966) reported a similar result in the intertidal, where the diversity of prey species is reduced if the predatory starfish *Pisaster* is systematically excluded. The implication is that competitive exclusion can only manifest itself when populations approach resource limitation and experience high levels of interspecific competition.

On the other hand, heavy cropping or predation may sway the competitive balance in favor of a reduced set of species that are able to withstand the pressure. This is observed in the tall-grass prairies of west-central Minnesota, which are mowed on the more level expanses for "prairie hay." Such areas rapidly lose their complement of the forb species that are so common in natural grasslands there, and eventually show a much simpler species composition of almost exclusively grasses. These observations indicate that maximal diversity of prey species occurs at predator pressures intermediate between very low and very high cropping rates. Dr. Kimball Harper has kindly lent unpublished results which confirm this picture. A series of open aspen (*Populus*) woodlands in Utah were arranged in order of increasing grazing stress (Figure 65);

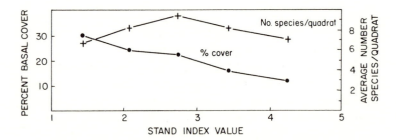

FIGURE 65. Relation between predation pressure and species diversity. Although plant cover decreases steadily with increasing grazing pressure, plant species diversity reaches a maximum at intermediate grazing pressure, where neither interspecific competition nor grazers depress it. From K. Harper, MS.

the percentage of basal cover decreased as this grazing pressure increased. The number of species per quadrat, however, was maximal at intermediate grazing levels and decreased toward both high and low extremes. Species were presumably eliminated by competitive exclusion when grazing was minimal, and the less well adapted of them eliminated by the herbivores themselves where cropping rates are high or "abusive."

II. BEHAVIORAL MODIFICATIONS OF
DISPLACEMENT PATTERNS

A. Interspecific Territoriality

1. The Relation Between Ecological Similarity and α_H. The single, most effective way to separate entirely the ecological activities of individuals is territoriality. The great majority of passerine bird species that have been studied advertise to neighboring individuals by voice and/or appearance their possession of a delimited area. This area becomes the territory, and its possession in the standard case entitles the owner to exclusive use, within that species, of the resources it contains. These are normally male-male interactions; inclusive resources are then mates, food, nest and shelter sites. Thus individuals of a single species, which are identical with regard to resource requirements, are completely separate in space: $\alpha_H = 0$ and $\alpha_F = \alpha_V = 1$, in this restricted sense.

The pattern which now emerges is that individuals of one, two, or more species which are extremely close in most ecological requirements do not tolerate spatial overlap in territories; individuals of different but ecologically similar species overlap horizontally (co-occur at a point in the habitat) but are separated in some other dimension (such as a vertical separation of feeding zones); and individuals of quite dissimilar species overlap in space both horizontally and vertically, but feed on different food items sought with different foraging techniques.

It is a logical extension of territoriality systems that pairs or larger sets of species which are ecologically similar may defend interspecific as well as intraspecific territories. Very often such species pairs are geographic replacements with only marginal sympatry ("stasipatry") or are species with distinct habitat preferences which meet in intermediate habitat.

Orians and Willson (1964) were the first to assemble cases of interspecific territoriality between bird species, and identified the following three situations in which the phenomenon might be expected: a) when the structure of the vegetation is simple, b) when species are adapted to exploit stratified food sources, and c) when other species are already exploiting similar resources.

I have further generalized these conditions (1969) to the following three situations in which the usual options in inter-species resource division may not be exercised: a') the habitat is simple (implying a narrow span of resource types); b') the species involved are food specialists (in Wrightian terms, a food specialist occupies the top of a steep-sided adaptive peak, and any compromise in resource use, such as its division with a second species, will severely reduce fitness; only when adaptive peaks are broad may more than one species reside on them); c') many other species are present (thus further restriction of utilization curves on the resource gradient is not feasible, except by elimination of one of the species already coexisting on the resource span). A stable coexistence of two species in a zone of marginal overlap implies that the resource or niche space can be stretched, or alternatively that niches can be contracted and still permit the persistence of all species involved. This increased species packing may not be possible in cases where, for historical, chance, or other reasons, maximal species packing already occurs. Tropical locations are possible candidates a priori for such type (c') situations, as are the centers of radiation or speciation for particular taxonomic groups (wrens in the neotropics, shrikes in Africa). One can readily find examples of interspecifically territorial pairs of species which each appears to fit one of these conditions. Recently, Murray (1971) attempted to argue that interspecific territoriality is nonadaptive. Such thinking stems primarily from an unreasonable blanket application of the "competitive exclusion" principle, and here belies much evidence to the contrary, in particular the evolution of character convergence discussed later.

2. A Quantitative Argument. It is useful to ask the question: At what level of ecological similarity will selection favor nonoverlapping territories between different species? Clearly this will occur when the benefits of interspecific territoriality—a more concentrated food supply spread over a smaller area—just more than offset the advantages of a wider foraging area. This can be seen in the following argument, which depends for its validity on the premise that individuals spend time necessarily but unproductively in traveling within the territory.

Territorial birds spend much of their time feeding their young, time which has two components, actual foraging time and time spent traveling between nest and feeding site. This latter is called "traveling time," and time thus spent is a factor of economic importance in bird time budgets. Three kinds of evidence support this contention: a) territories are often rounded as opposed to elongate in uniform habitat, and b) nests are often positioned close to the territory center. Further, c) the young of species which travel long distances with food grow more slowly than those of similar species with light traveling requirements, and are disadvantaged thereby.

Figure 66 uses data from Craighead and Craighead (1956) to plot territory size against maximum territory width in hawks and owls. Only when territories are polygonal are the requirements of both maximum space utilization and best roundedness of the territories satisfied; such a division is represented by the lower boundary in Figure 66. It can be seen that deviations above this line are not extensive, especially for larger territory size when traveling time might be more important. In some species territories are actually or very closely polyhedral in shape (Grant, 1968); this and other reports of approximately circular territories refer to species breeding in uniform tundra or grassland (Harris, 1944; Drury, 1961). Nice's (1943) song sparrow territories show the same con-

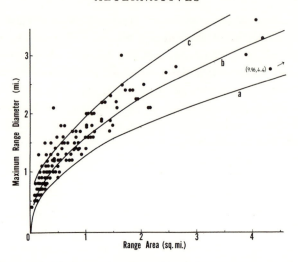

FIGURE 66. Maximum diameter of hawk foraging range versus area of foraging range, showing that foraging ranges tend to be rounded rather than elongate in shape. Curve *a* represents a circle, *b* an ellipse with major axis twice the minor axis, and c an ellipse with major axis three times the minor axis. Data from Craighead and Craighead, 1956.

sistency of shape, even though prime habitat (in this case bushes along ditches) is noncontinuous and linear in its distribution. The second point is illustrated by maps of territory and nest position such as those of Pitelka, Tromich, and Treichel (1955) for jaegers, of Cade (1960) for falcons, and for the longspurs *Calcarius ornatus* and *C. lapponicus* of Harris and Drury (*op. cit.*); nests are at or remarkably close to the center of the territory. Lastly Cody (1973) found that, in the six-species alcid communities studied in Washington, the growth rates of the chicks is inversely related to the distances the parents must fly to find food for the young. The existence of selection for shortening the fledging period as much as possible is supported by many general references, and from observations which confirm that the heaviest nestling mortality occurs in chicks of the alcid species (Cassin's auklet) feeding farthest from the nest site.

We assume that natural selection favors production of the largest possible number of young, and that food is limiting. The number of young which can be reared by a pair of birds will thus be proportional to the food uptake per day of the parent birds, which is in turn proportional to (1) the fraction of their time spent foraging and (2) the abundance of food. The number of young must also be affected by (3) the size of the area over which the birds can forage; while there must be a law of diminishing returns for larger territory size, I assume that the population is sufficiently dense that the two are directly proportional. Representing the number of young by n, we have

$$n = kt_f FA = k(p - t_t)FA,$$

where A is territory size, F is food abundance, and p the proportion of time parents spend feeding young, $=$ traveling time $t_t +$ foraging time t_f. The k is a constant of proportionality. In the first case we let two competing species maintain completely overlapping territories. The expression

$$n_1 = k'r^2F(p - k''r/\sqrt{2})$$

gives the number of young one species can raise simultaneously in the presence of the other species. Territory size A is written πr^2 and π is absorbed into the constant k'. Traveling time is approximated in proportion to the radius of a circle enclosing half the area of the territory, i.e. proportional to the distance traveled in an average foray if the territory is uniformly searched $(A/2 = \pi(r/\sqrt{2})^2)$. The k'' is a second constant of proportionality. A similar expression describes the number of young raised by the second species. Now suppose these two species are interspecifically territorial (no territorial overlap between them). That is, the same number of pairs of each species is present in the habitat, but now the size of their territories is reduced by half. Then, A becomes $A/2$, and the radius of a circle enclosing half the new territory size is $r/2$ $(A/4 = \pi(r/2)^2)$. Food resources now cover a smaller area, but are increased in density by a factor $(1 + \alpha)$, where α is an index of niche overlap or competitive similarity. Thus traveling time is reduced, but this economic advantage may be offset by a reduction in the total food available to $A(1 + \alpha)/2$. The number of young which can be raised under these new circumstances is

$$n_2 = k'r^2(1 + \alpha)F(p - k''r/2)/2$$
$$= n_1\left[\frac{(1 + \alpha)(p - k''r/2)}{2(p - k''r/\sqrt{2})}\right].$$

If n_2 is greater than n_1 there is an advantage to the species in interspecific territoriality. The condition is satisfied when the term in squared brackets is greater than unity. Note that, when $\alpha = 1$, this term is greater than unity for any value of r, but that increasingly smaller values of α require increasingly larger values of r for this to be so. More precisely, write n_2/n_1 as N and p/k'' as K to obtain

$$N = (K - r/2)(1 + \alpha)/2(K - r/\sqrt{2})$$

as plotted in Figure 67. The axis $N = 1$ shows the value for which no advantage or disadvantage results from interspecific territoriality. The r-α plane is divided into two zones by the line $N = 1$, which has the equation

$$r = K(1 - \alpha)/(1 - \alpha/2).$$

The intercepts of this "isocline" on the ordinate α and abscissa r in Figure 68, which illustrates this, are $(0,1)$ and approximately $(K,0)$. Pairs of species with α and r values between the isocline and the origin derive no advantage from interspecific territoriality, whereas species with values beyond this line do. Figure 69 shows a plot of α vs. r, using all examples of (breeding) interspecifically territorial pairs of species cited in Orians and Willson (1964), for which territory size is given

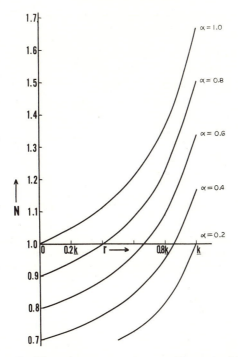

FIGURE 67. The advantage to interspecific territoriality (N, positive when N exceeds unity) as a function of territory radius r and an index of ecological similarity α. For reduced values of α between species, only very large territories should be defended interspecifically.

in the original reference. An unpublished observation by Cody of interspecific territoriality between Baird's and grasshopper sparrows (*Ammodramus bairdii* and *A. savannarum*), where their ranges overlap in southern Saskatchewan, is included, as is later work on finches (Cody and Brown, 1970; see below) and Tyrannids (see below). A logarithmic scaling of r is used to produce a correspondence to Figure 68. Thus K is not a constant, but a function of r, and increases as does r. Now $K = p/k''$; it is unlikely that p increases with larger territories, more reasonably p would decrease due to larger defense

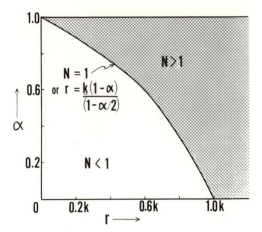

FIGURE 68. Pairs of species should defend interspecific territories only if their territories (radius *r*) and ecological similarity *α* place them in the crosshatched part of the *α-r* plane.

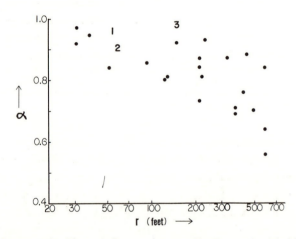

FIGURE 69. Plot of known interspecifically territorial bird species. Ecological similarity *α* is estimated by the ratio of bill lengths. Dots are species pairs cited in Orians and Wilson, 1964. Additional points are 1) *Pipilo ocai* and *Atlapetes brunneinucha*, 2) *P. ocai* and *P. erythrophthalmus*, 3) *Ammodramus savannarum* and *A. bairdii*.

222

times for larger areas. More likely k'' is decreasing with increasing r, perhaps reflecting a nonuniform use of larger territories (more feeding activity toward the center).

3. *Intraspecific and Interspecific Territoriality in Time.* We can conceive of territorial defense over time just as easily as over space. Individuals might simultaneously occupy partially or completely overlapping ranges in space, but maintain a temporal segregation within these ranges in the use of this space. A regular or irregular diurnal or other periodic cycling of activities related to resource use could permit the coexistence of individuals within or among species. Documentation of such behavior, however, is quite sparse.

Logically we reason that resources best defended by temporal territoriality would be those that are a) renewable at a relatively rapid rate, b) perhaps also widely scattered, or both. Thus temporal territoriality appears to be an alternative to coloniality or social use of resources. The nectar supplies of flowering plants (or the insects attracted by them) are an example of such food resources, for either the nectaries are replenished or additional flowers mature continually through time. Hummingbirds, the chief avian users of these resources in the western hemisphere, in fact use a combination of spatial and temporal territoriality. There is a rapid shifting of territories and replacement of (temporary) owners around flowering bushes in the Mohave Desert, both within and among species (Cody, 1968b). Moynihan (1963) described a temporal segregation among the four species of nectivorous birds (Coeribidae) using flowers in the Columbian Andes. The maintenance of territorial separation in hummingbirds is obviously accomplished by the visual signals of the bright-plumaged males; even the duller females are iridescent. The Andean birds, however, are dull in appearance, and segregation is apparently maintained by a steady vocalization.

The mechanism of temporal territoriality may be quite

widespread both within and among species, but, of course, is much more difficult to detect than conventional spatial territoriality. It may be in operation, for example, in California chaparral to facilitate coexistence between the two towhees *Pipilo erythrophthalmus* and *P. fuscus*. These two species show an extremely high value of ecological similarity, the highest in the community ($a = 0.830$, compared to the community mean of 0.566), but this index takes no account of temporal factors affecting niche overlap. An index of habitat preference overlap between the species within the study area, α_{H1}, shows a value of 0.918, and likewise their territories are largely overlapping ($\alpha_{H2} = 0.848$); hence no apparent interaction between them occurs to reduce spatial overlap. The food supply for both is ground and leaf-litter insects and seeds, obtained similarly in both species ($\alpha_F = 0.841$) by scratching and searching. It is conceivable that they minimize coincident use of a habitat patch by maintaining a separation in time, using vocalizations to inform each other of their whereabouts in the habitat. The only evidence for this is a) the qualitative observation of noncoincident use of habitat patches in the two species, and b) the observation that their vocalizations tend to be synchronized in time. Brown and Cody collected data (other results of which, including methods of observation and analysis, were published in 1969) that indicate a weak positive correlation between song peaks in spotted and brown towhees in lowland chaparral (two mornings) or no correlation (one morning). Clearly more work is required to resolve this point for the towhees, and it appears to be particularly promising: only one of the two species, *P. erythrophthalmus*, occurs above 7000′ in the Angeles National Forest, and the other is apparently "squeezed out." Again, in high-elevation successional chaparral this towhee niche supports only one "ecological" species, although two taxonomic species co-occur there (see below the Blue Ridge finches, II.B.2). It is conceivable that a similar mechanism of temporal segregation to that postulated here accounts for the coexistence and territory overlap between the very similar *Pipilo fuscus* and *P. aberti* in Arizona (Marshall, 1960), and perhaps also for the unexpected vocal similarities between the cardinal *Richmondena cardinalis* and *Pyrrhuloxia sinuata* in Arizona (Lemon and Herzog, 1969), which are likewise ecologically similar but spatially overlapping.

The hypothetical temporal patterns discussed here are tantamount to a maintenance of individual distance in space, but which allows the actual positions of the individuals to vary over space in time. I watched lesser nighthawks forage over a quadrangle on the University of Houston campus in summer 1971. These birds hawked at dusk for flying insects, and were remarkably well spaced out horizontally (mostly between 100' and 200' above the ground), so that only very rarely were more than two or three birds above the four acre quadrangle at any one time, although they were moving rapidly and apparently erratically over this and adjacent areas. Different individuals came and went every few seconds, but expected chance accumulations never occurred. Every three or four seconds the birds gave their characteristic calls which I suggest serve to maintain their individual distances in the air, and to prevent successive searches of the same air space within short time intervals. The time interval between successive nighthawk visits should be related to the rate at which insects reoccupy (from above or below) searched and harvested air space. Indications which tend to confirm this are a) the increased frequency—and pitch—of the calls when two individuals approach each other, and a resultant veering off in their courses, and b) the much reduced frequency of calling in lone individuals (once per 10 or more seconds) foraging over presumably less productive habitat in which chance pile-ups would be less likely.

Temporal territoriality is not as conveniently maintained by voice (nor, of course, by visual display) as by scent. Odors might be evolved with fading rates, or behavioral threshold levels to odors might be selected, such that schedules of foraging visits are appropriate to the renewal rates of the resources. Many mammals, for example, are not territorial in the way most birds are, but maintain home ranges that are not defended except around the den and that overlap those of other individuals. Scent signals predominate among the Felidae, Canidae, and Mustelidae, whose food supplies, primarily mobile prey items, are continually renewing in the habitat. Conceivably the use of habitat within these overlapping home ranges might be regulated by a scent-oriented mechanism of temporal territoriality. This is apparently just what happens among feral cats in Germany, according to Leyhausen and Wolff (1959). Several individuals made use of the same hunting areas, but their activities were spaced in time so that no conflicts resulted. Further studies of other likely scent-oriented mammals are required before the generality of time-coordinated use of space is established.

B. Partial Interspecific Territoriality

1. *The Intermediate Situation.* It is apparent that interspecific territoriality may not be an all-or-none phenomenon, and that intermediate cases may occur in which ecologically rather less similar species exhibit only partial territorial overlap. Such interactions might be quite subtle in that little or nothing in the way of overt aggression or convergently similar territorial-defense signals occurs. Two species might simply evolve a mutual avoidance reaction to each other's originally intraspecies spacing signals; such resultant "interactions" are both detectable and measurable.

One way of showing the existence of such subtle interspecies spacing is to contrast two ways of measuring interspecific horizontal overlap α_H. Two species that co-occur in, say, the same ten-acre patch will each occur at certain grid points not covered by the other, numbering p_{11} and p_{22}, and will each hold some points in common $-p_{12}$. The spatial overlap is measured by the formula $\alpha_{H1} = p_{12}/((p_{11} + p_{12})(p_{22} + p_{12}))^{1/2}$, by convention and avoiding questions of niche breadth. But now if at each grid point the habitat (vegetational) characteristics important in habitat selection are measured (e.g. vegetation height, vegetation density), then the two species may be shown as clusters of points in n-space, if n such habitat variables are included. Their overlap in this space (or plane, if, for example, the two parameters above are used) can be measured exactly as before to give α_{H2}. Notice that $\alpha_{H2} > \alpha_{H1}$, but that if there is no interaction between the two species $\alpha_{H1} = \alpha_{H2}$. Thus the deviation from equality $(\alpha_{H2} - \alpha_{H1})$ measures the strength of the interaction between the two species, and varies from 0 (no interaction) to 1 (the interaction customary within in a species, with no difference in habitat among territorial pairs but no overlap in space.[1]

As an illustration of the sort of interspecific interaction inferred by $(\alpha_{H2} - \alpha_{H1}) > 0$, consider the finches of the grass-sage site in Wyoming. Four species are present, Brewer's sparrow, white-crowned sparrow, vesper sparrow, and savannah sparrow, which display between pairs rather low overlaps in space. Table 16 contrasts observed overlap in vegetational characteristics of the finch habitats, including the number, height, and density of the sagebrush individuals, with that in territorial disposition among them. Between pairs of the first three species mentioned, $(\alpha_{H2} - \alpha_{H1})$ is close to zero, but be-

---

[1] We shall exclude from analysis, for lack of information, the possibility that two species might have discrete and unrelated (and distributionally uncorrelated) requirements of their common habitat. Consider a species which forages only in 3′ high bushes, and another which forages only in 4′ high bushes also over 5 acres, 2½ of which are included in the first species range and therefore support both 3′ and 4′ bushes, $\alpha_{H1} = 0.50$, and by my habitat measuring techniques $\alpha_{H2} = 0.50$ also. But in fact the habitat requirements are potentially quite independent for the two species, and α_{H2} would be zero if we knew exactly what to measure.

TABLE 16. Interactions between the finches of Wyoming grassland-sagebrush habitat. The first figure represents overlap in territories, the second overlap in the vegetational characteristics of the same territories. Large differences between the pairs of overlap values represent interactions between the species pairs. B = Brewer's sparrow, V = vesper sparrow, S = savannah sparrow and W = white-crowned sparrow.

	B	V	S	W
B	1.000/1.000	0.868/0.873	0 /0.403	0.588/0.645
V		1.000/1.000	0.009/0.493	0.591/0.591
S			1.000/1.000	0 /0.523
W				1.000/1.000

tween these species and the savannah sparrow a significant interaction occurs, with values for the difference between the two overlap measures of 0.403, 0.484, and 0.523. Thus it appears that savannah sparrows tolerate little spatial overlap with the other finches with which they co-occur, but exclude them (or vice versa) from territories. It might be added that two of the other three sparrow species habitually occur and overlap in territories with other species (e.g., Brewer's sparrow in Colorado grass-saltbush, and white-crowned sparrow in the Teton willows), but the savannah sparrow in Minnesota grassland shows almost complete habitat (and spatial) separation with the other two species present (bobolink and LeConte's sparrow) (Figure 6), and indeed with other grassland birds in the vicinity (sharp-tailed sparrow, grasshopper sparrow, meadowlark). In spite of the magnitude of this apparent interaction, no overt aggression between savannah sparrow and the other species was observed. Apparently rather subtle avoidance reactions suffice to segregate species.

Before discussing further examples of these behavioral interactions and their role in community organization, we should provide a word of caution. A possible source of error is that a species, in the year its territories were mapped, did not completely fill up the available habitats it might otherwise have done had its numbers been higher. Heavy overwinter mortality or poor reproduction in the previous season or both may contribute to an unusually low density of the breeding population. Then preferred habitats will fill up first, and α_{H1} will be underestimated. There are two ways around this: to visit sites over several seasons so that a reasonably dependable estimate of population density and habitat use is obtained, or to measure preferred habitat sufficiently accurately that preferred and less-preferred habitats are distinguished. If the latter can be done, this source of error is then absorbed by the second possible source, which comes about through inaccurate identification by the observer of habitat variables on which birds base their choice of breeding territory. This possibility, which might overestimate α_{H2}, is minimized by selection for analysis of those structural attributes (height, density) of the vegetation which previous studies on species diversity and niche relationships have labeled as important. They are then combined in such a way, for instance by discriminant functions, that each contributes maximally to the separation of species in the multidimensional habitat space. Lastly, if the apparent interactions between species pairs make ecological sense—if they occur between species, for instance, which are similar in feeding behavior and foraging height, and possess a similar bill morphology—then confidence in the measure of the interaction is justified.

Further illustrations of the relation between overlap in space and overlap in vegetational characteristics of the habitat are shown in Figure 70: a) Points might fall immediately under the 45° line, indicating both accurate measurement of vegetational characteristics important to the birds and a lack of behavioral interactions between species; b) the points might lie parallel to the 45° line but somewhat displaced down from it; this is more likely to reflect a consistent overestimate of α_{H2} rather than an interaction of consistent size between a great variety of species pairs; and c) points may tend to di-

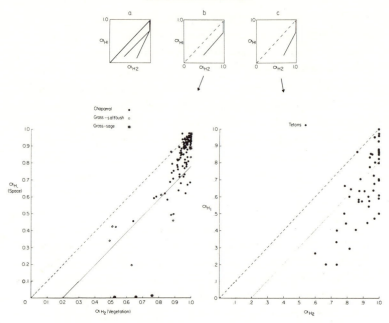

FIGURE 70. Habitat overlap can be measured in two ways, as a simple spatial overlap α_{H1} or as overlap in the vegetational characteristics of two species territories α_{H2}. α_{H2} deviates above α_{H1} to the extent to which the species interact with each other behaviorally to accomplish greater segregation. Bird species pairs in chaparral do not deviate in the two measures by more than 20%, which almost certainly represents not a behavioral interaction but inaccurate reading of appropriate vegetational characteristics. In the Teton Willows, behavioral interactions become stronger as habitat is increasingly used as a major means of ecological segregation.

verge away from the 45° line with decreasing α_{H1} and (less rapidly) decreasing α_{H2} values. This trend is observed in both Wyoming sites and the Colorado site, in each of which segregation by habitat is well-marked. It is absent from the chaparral site, in which extensive habitat overlap is the rule. The data indicate that, as habitat segregation becomes a more and more important part of interspecies ecological segregation, the species involved are increasingly likely to supplement differential habitat selection by interactions that further decrease their actual spatial overlap below habitat overlap.

2. *The Blue Ridge Finches.* The Blue Ridge extends southeast from the main east-west ridge of the San Gabriel Mountains in southern California, at around 8400′ elevation, and overlooks Los Angeles and its eastern satellite cities. The climax vegetation is chiefly jeffrey pine (*Pinus jeffreyi*) and white fir (*Abies concolor*), but the southern side of the ridge burned in 1953, and is under a successional chaparral with scattered pines. This chaparral is composed of the same plant genera (*Ceanothus, Arctostaphylos, Ribes*) as the putative climax chaparral of lower elevations from sea-level up to perhaps 7000′, with which it is continuous. In spite of this continuity, almost none of the 20–30 bird species of the lower elevation chaparral breeds in this habitat, and instead its avifauna is drawn from locally adjacent sagebrush, coniferous forest, and streamside willows.

In Table 17 the birds of Blue Ridge chaparral are compared with those of structurally similar habitat in the Santa Monica Mountains, at around 2000′ on a similar south-facing ridgetop. In species number, the slight difference is chiefly in the absence of more foliage insectivores at high elevation. In species names, however, almost a complete turnover occurs. Replacements are within the same genus in six cases, within the same family in eight cases, and across families in four cases. Only three species are in common between the two censuses: red-tailed hawk, a really cosmopolitan species; red-shafted flicker, another widely-distributed bird; and the black-chinned sparrow. Because of the general similarity in taxonomy between species replacements, these are matched with facility.

Of course, it is a common circumstance that within-genus and also between-genus replacements occur along habitat gradients. But here the structural similarity between the two

TABLE 17. Comparison of breeding birds in low-elevation climax chaparral and high-elevation successional chaparral, San Gabriel Mts., California.

Blue ridge, 8400'	South slope, 2000'
Foliage insectivores	
	Wrentit
	Bushtit
House wren	Bewick's wren
Mountain chickadee	Plain titmouse
Sallying flycatchers	
Hammond's flycatcher	Western flycatcher
Townsend's solitaire	Ash-throated flycatcher
Nectivores	
Anna's hummingbird	Anna's hummingbird
Calliope hummingbird	Costa's hummingbird
Ground feeders	
Mountain quail	California quail
Band-tailed pigeon	Mourning dove
Green-tailed towhee	Brown towhee
Fox sparrow	Spotted towhee
Black-chinned sparrow	Black-chinned sparrow
Chipping sparrow	Rufous-crowned sparrow
Oregon junco	Sage sparrow
Stellar's jay	Scrub jay
Purple finch	House finch
Mountain bluebird	Black-headed grosbeak
Robin	California thrasher
Woodpeckers	
White-headed woodpecker	Nuttall's woodpecker
Red-shafted flicker	Red-shafted flicker
Aerial feeders	
Violet-green swallow	White-throated swift
Raptors	
Goshawk	Cooper's hawk
Red-tailed hawk	Red-tailed hawk

sites compared is great. There is, however, a very considerable difference in altitude, and perhaps associated with this a difference in the availability of food resources through time. Yet it is difficult to avoid the conclusion that differences in *physiological tolerance* between congeners in particular are chiefly responsible for the species turnover. I do not know of comparable evidence that is quite so convincing; here habitat structure and therefore the spatial dispersion of food remain very similar, and yet the climatic environment changes drastically to favor apparently the utilization of these food resources by different species.

The role of the large ground-feeding litter-scratching finches, the "towhee niche," is taken up by *Pipilo erythrophthalmus,* the spotted towhee, and *P. fuscus,* the brown towhee, from sea-level to at least 5000′. Above this elevation the brown towhee drops out, perhaps excluded in a diminishing resource space by competition with the spotted towhee. At 8000′ the spotted towhee occurs only in willow thickets. On the Blue Ridge the towhee niche is co-occupied by the fox sparrow *Passerella iliaca* and the green-tailed towhee *Chlorura chlorura* (both $6\frac{1}{4}''$ in body size), the former being recruited from the willows and the latter from the sagebrush. Both species were common in the 11.65 acres of successional chaparral mapped on the ridge; the towhee occurred in 82 of 203 $50′ \times 50′$ squares censused, the fox sparrow in 108. The interrelations of these two species will be discussed below.

A second species group comprises three finch species. The lowland species black-chinned sparrow, rufous-crowned sparrow, and sage sparrow are quite similar in ecology and in body size ($5\frac{1}{4}''$, $5\frac{1}{4}''$, $5''$, respectively); they are characterized by similar feeding height distributions divided between the ground and the lower bush foliage. On Blue Ridge two of these three are replaced by chipping sparrow and Oregon junco, and the black-chinned sparrow is retained ($4\frac{3}{4}''$, $5\frac{1}{4}''$, $5\frac{1}{4}''$). While the trio coexists apparently quite comfortably in lower chaparral, it does not at 8400′, as is discussed below.

The five common finches on Blue Ridge, the ecologically similar duo green-tailed towhee and fox sparrow, and the trio junco, chipping and black-chinned sparrows, quite obviously have rather different habitat preferences. The territories of all five were mapped in May–July 1971, and the occupancy or otherwise by each species of each 50′ × 50′ patch of the study was determined. Each patch was also characterized by its cover by (understory) chaparral of height 0–1′, 1–3′, 3–5′ and >5′ in height, and its support of numbers and total height of larger pines and firs that escaped the fire or grew up since 1953. The four variables most obviously involved in habitat segregation—a) proportion of patch covered with brush less than 1′ high, b) proportion of patch covered with brush greater than 3′ high, c) number of trees in the patch, and d) total height of all trees in the patch—were entered as variables in a discriminant function analysis, and from the resultant species distributions along this linear combination of the habitat variables the habitat overlap between species pairs was calculated. This overlap, α_{H2}, is the figure to the right in the matrix of Table 18. A better idea of the habitat relations among the five finches is obtained from Figure 71, which shows 99.9% confidence or contour ellipses for each

TABLE 18. Interactions between finches on Blue Ridge, San Gabriel Mountains, southern California. Figure to the left of slash is overlap in territories, or spatial overlap, figure to the right is overlap in vegetational characteristics of the same territories. F = fox sparrow, G = green-tailed towhee, B = black-chinned sparrow, C = chipping sparrow, J = Oregon junco.

	F	G	B	C	J
F	1.000/1.000	0.341/0.658	0.426/0.472	0.253/0.379	0.360/0.456
G		1.000/1.000	0.282/0.431	0.348/0.500	0.396/0.603
B			1.000/1.000	0 /0.511	0.038/0.457
C				1.000/1.000	0.265/0.761
J					1.000/1.000

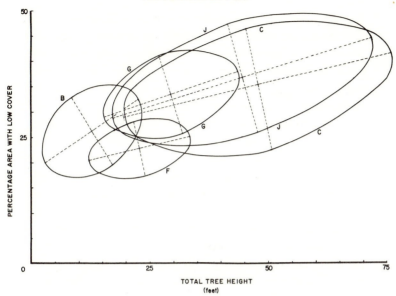

FIGURE 71. Habitat preferences of five finches in successional chaparral on Blue Ridge, San Gabriel Mountains, southern California. B = black-chinned sparrow, F = fox sparrow, G = green-tailed towhee, C = chipping sparrow and J = Oregon junco. Ellipses are 90% confidence limits of habitat distribution for each species.

species in the two habitat variables, (a) and (d), which figure most prominently in the discriminant function; black-chinned sparrow is at the tall brush-no trees end of the habitat gradient; and chipping sparrow is at the no brush-many trees extreme. In fact the species are almost linearly aligned in habitat preference, for Mahalanobis' D^2 statistic, calculated between each pair of species can be used to arrange species along a line with but little ambiguity (chipping sparrow, junco, green-tailed towhee, fox sparrow, black-chinned sparrow).

Thus there are certainly differential habitat preferences which will separate the territories of the species in this habitat.

235

But when the territories of the five are actually plotted, their spatial separation (α_{H1}, left-hand figure of matrix entries of Table 18) is in several cases far greater than the differences in habitat preferences would predict. Furthermore, ($\alpha_{H2} - \alpha_{H1}$) is great between just those species already labeled ecologically similar, the two large litter-scratching species and the three smaller ground and low brush finches. The size of the interaction averages 0.438 ($n = 4$) between species in the two groups of species expected to interact, and only 0.129 ($n = 6$) between pairs across the two groups.

There is also direct evidence for a behavioral interaction between the fox sparrow and the green tailed towhee, and the mechanism responsible for this is apparently the evolution of similarities in their songs. Normally one has no difficulty in distinguishing the songs of these two species, for the towhee's is shorter with a long trill, the sparrow's a long, fluty melodic jumble of notes. Although some songs of each species on Blue Ridge are the typical sort, some are not, and even with practice the identity of the vocalist may be confused. Thus there seems to have evolved an intermediate song, which involves interspecies interactions in a very demonstrable way: a) Initiation of song in one male stimulates neighboring males of both species to sing; b) neighboring males of different species engage in "song duels," in which the singing individuals alternate quite precisely in delivering songs; c) songs of one species, when played back in the territory of the other, stimulate the territory owner to approach with great agitation to within a few feet of the recorder and to sing vigorously; d) the two species actually fight, and antagonistic encounters take place between species (chases and threat attacks, scuffles occur commonly as within species).

The songs of fox sparrows and green-tailed towhees from Blue Ridge are shown in Figure 72. These range from the quite different "typical" or "normal" song in each to an intermediate version. Of further interest is the tendency in each

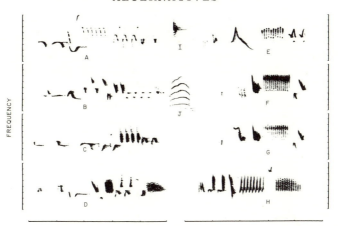

FIGURE 72. Sonograms of fox sparrows (left) and green-tailed towhees (right). A. Typical fox sparrow, from R. T. Peterson record. B. Typical fox sparrow on Blue Ridge. C and D. Blue Ridge fox sparrows "intermediate" between typical fox sparrow and typical green-tailed towhee. E. Typical green-tailed towhee from R. T. Peterson record. F. Typical green-tailed towhee from Blue Ridge. G and H. Green-tailed towhees from Blue Ridge "intermediate" between typical green-tailed towhees and typical fox sparrows. I. Fox sparrow alarm call. J. Green-tailed towhee alarm call.

of the two species to alternate in their song sequence a typical and an intermediate song, but the further studies which are in progress will elucidate this. It is perhaps also of significance that the call notes of the two (Figure 72) are dissimilar and unvariable (see II.C.2 below). In Figure 73 the songs of chipping sparrow, Oregon junco, and black-chinned sparrow are given, and it seems likely that the same mechanism as in the sparrow-towhee case, response to similar songs across species, is responsible for their spatial separation. But unlike the other two, the trio have little or no intraspecific song variation in this locale. Chipping sparrow songs appear to the ear to be unvariable, junco songs show only slight variations, and

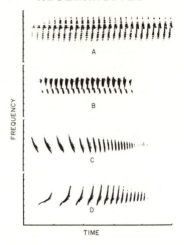

FREQUENCY

TIME

FIGURE 73. A. Chipping sparrow song. B. Oregon junco song.
C and D. Black-chinned sparrow songs, usually alternated.
All from Blue Ridge.

black-chinned sparrows have two different songs which are
often alternated in a sequence. Of interest is the very basic
similarity between all three. Each has a song which consists
of a single note, and each trills this note for about 2½
seconds. The note is rather different in structure between the
chipping sparrow and junco, but repeated at the same pitch,
and in black-chinned sparrow the note is either slurred down
(like the chipping sparrow) or up (like the junco), but is
repeated at increasing rate. No more direct evidence is yet
available on their interspecific interactions, but it is expected
that further work with recorded songs will produce this.

C. Character Convergence

1. The Hypothesis. Consideration of partial interspecific territoriality in the Blue Ridge finches and discussion of the probable mechanism of its operation, song convergence, leads us
next to a treatment of the phenomenon of "character con-

238

vergence" (Moynihan, 1968; Cody, 1969). The hypothesis and term were produced independently in these two papers, but the emphasis in the former was on the evolution of convergent characteristics to facilitate gregariousness and in the latter to facilitate spacing by aggression. It is proposed here to restrict the term *character convergence* to the latter situation, which is more closely an opposite and alternative to character displacement, and reserve Moynihan's term *social mimicry* for the former phenomenon, discussed below (II.D).

The character convergence hypothesis states that interspecific territoriality is sometimes associated, and presumably developed in parallel with, a convergence between species in those characteristics used to defend the territory, appearance and voice. In territorial defense, males would thus fail to distinguish between other males of their own and males of the convergently similar species, and the two would divide space as a single species. Interbreeding is prevented by the retention of species-specific recognition cues used by females to select mates of their own species (mate selection is usually the female's choice in birds; Orians, 1969). Attempted hybridization and its attendant ecological disadvantages is thereby avoided.

Character convergence may evolve wherever conditions favor interspecific territoriality, as discussed above. As the advantages in spatial separation in ecologically similar species are mutual, evolution may proceed simultaneously in both species toward new convergent phenotypes. Alternatively, and perhaps more usually (see the Chilean bird species discussed below), it may be a one-sided change. A mutant individual of species A which is more similar in territory defense signals to species B will be partially excluded from the territories of the latter. Such a mutant is only favored if either a) it is now forced to breed in habitat not occupied by B, a less interesting alternative but one which must have occurred many times in the evolution of precise habitat selection, or b) it

239

still coexists in the same habitat as species B, by reciprocating the aggressive behavior B displays toward it. In this latter case, A enjoys, from the argument presented above, greater reproductive success than its conspecific neighbors which overlap in territory with B, and the new phenotype comes to predominate in the population. The convergent characteristics may then be further selected until complete territorial exclusion between species A and B results. At one level, then, displacement patterns in habitat use are developed among ecologically similar species to increasingly fine distinctions; beyond this, habitat co-occupancy is possible only through the medium of interspecific territoriality.

That such an arrangement has been evolved many times in many different animal groups is shown below. It is only surprising that interspecific territoriality apparently occurs often with no such convergence between participants. Notably, however, these cases almost always involve congeners (Orians and Willson, 1964), except for owls and hummingbirds (but see Cody, 1968b, 1969, on the lack of intergeneric divergence in female trochilids), and thus it is difficult to distinguish convergence from genetic affiliation. For this reason, and the fact that ecologically similar species in the same habitat are likely to be *generally* similar in voice and appearance, being products of the same environments, we are likely to identify only the most obvious examples of character convergence.

2. The Evidence. Both appearance and vocalization may provide the signals by which birds defend territories. First we treat convergence in appearance.

Two species groups provide only circumstantial evidence. These are southeast Asian woodpeckers Picidae and African bush-shrikes *Malaconotus* and *Chlorophoneus*. The woodpeckers include several pairs of genera, such that one genus per convergent species pair comes from each of two taxonomically distinct sections of the family. Sympatric species in these distinct genera show remarkable similarities in appearance. Most striking are the species of *Dynopium* (*javense* and *benghalense*) and *Chrysocolaptes* (*lucidus*), difficult to separate in the field in spite of their divergent taxonomy (Plate 1 of Cody, 1969). The convergence in appearance extends to coincidence of coloration—red, yellow, black, and white—and patterning, which varies clinally in parallel across the genera from north to south in India and Ceylon. The only subspecies which are distinct (not convergent) are those in allopatry: *D.b. dilutum* and *D.b. benghalense* (part) in northwest India and West Pakistan, and six subspecies of *C. lucidus* on the Philippines. The characters which remain different in the convergent species are malar stripe and bill color—both associated by limited evidence with mate selection in the Picidae—and voice. Thus everywhere *Dynopium* has a dark bill and *Chrysocolaptes* a pale one except in some allopatric subspecies.

The natural history of these species is little known, and no direct evidence for their interspecific territoriality exists. Observations on a greater abundance of *lucidus* and more extended habitat use where it occurs in allopatry than where it is sympatric with *benghalense* imply a competitive interaction; it is predictable that behavioral interactions will spatially separate the species where they meet on intermediate ground between their respective, preferred habitats of moist and dense versus drier and more open forests.

The shrikes provide a similar picture, with quite exact similarities in appearance linking pairs of species, one from each of the genera *Malaconotus* and *Chlorophoneus* (color plate in Hall *et al.*, 1966, who propose merging the two genera into *Malaconotus*). The convergently similar pairs are those which occur together in a habitat type, and the species within each genus are separated by habitat. Thus savannah, rain forest, and montane forest each support a pair of taxonomically distinct (at a level approaching the genus, at least) but visually

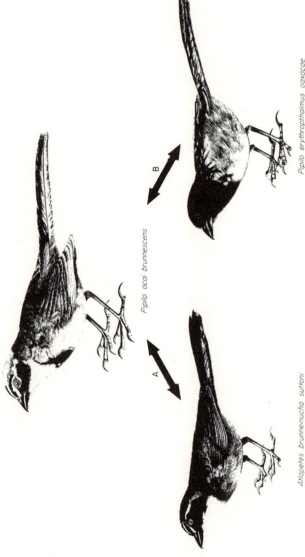

Pipilo ocai brunnescens

Pipilo erythropthalmus oaxacae

Atlapetes brunneinucha suttoni

B

A

Atlapetes pileatus pileatus

similar species. In the savannah, indirect evidence associates one of the pair with more open and the other with less open habitat, and although they are widely, almost completely sympatric, most collecting localities produce only one of the pair. Voice and eye-color might provide the necessary species-specific recognition cues to mate selectors.

Convincing evidence of the association of like appearance with interspecific territoriality was provided by Cody and Brown (1970), involving again and perhaps not coincidentally towhees, *Pipilo*. *P. erythrophthalmus* extends over North America and south to South-central Mexico, where it meets the Mexican endemic *P. ocai* in brushy, forest-edge habitats. The two hybridize everywhere except where a third species, the brush-finch *Atlapetes brunneinucha*, is also present, on Cerro San Felipe in Oaxaca. *A. brunneinucha* and *P. ocai* are extremely similar in appearance (Figure 74) the more striking because of their bright green, chestnut, black, and white plumage. The two are interspecifically territorial on Cerro San Felipe, and behave ecologically as a single species. Only the former occurs within the forest, and only the latter on the drier, more open habitat lower on the mountain. Differences in voice may aid mate selection. We concluded that selection would act against *ocai-erythrophthalmus* hybridization by selecting against the *ocai*, which lost their *brunneinucha*-like appearance and hence their profitable interaction with the *Atlapetes*.

Atlapetes brunneinucha ranges widely through Central America to South America, and shows considerable geographic variability (Chapman, 1923), and elsewhere it has been found to occur with other ground-scratching finches with

FIGURE 74. Interspecific territoriality in Mexican finches. Species connected at A show a character convergence in appearance, the species at B a character convergence in voice. The fourth species is ecologically distinct and is not involved in interspecific relations. From Cody and Brown, 1970.

just the same colors and patterns. In Vera Cruz, for example, *brunneinucha* is found with three saltators, *Saltator atriceps*, *S. maximus*, and *S. coerulescens* and the sparrow *Arremon aurantiirostris* (Wetmore, 1943). At least two of the saltators vary geographically in some of the characteristics—e.g., throat color—in which they resemble the brush-finch. These finches, plus some others, may be involved in a character convergence complex which extends throughout Central America and into South America.

This is a convenient point to stress that unusual similarity in appearance in sympatric species which are not necessarily closely related appears to be a common phenomenon. In some cases the species vary geographically in parallel with each other, accenting species interdependence and inviting an explanation in terms of character convergence. Alfred Wallace (1869) came upon such a case on the island of Buru in the Malay Archipelago, where an oriole *Oriolus* (Oriolidae) and the friarbird *Philemon moluccensis* (Meliphagidae) look identical. On the nearby island of Ceram, the two appear quite different from the Buru birds (and are taxonomically distinguished), but again resemble each other quite precisely. This relationship between the two taxa extends to the New Guinea representatives and seems much more likely to have evolved in relation to interspecific aggression rather than to be a case of (Batesian) mimicry as postulated by Wallace.

During work in Peru, John Terborgh (pers. comm.) turned up several possibilities. Two insectivores of the leaf litter in deep forest, a wren *Microcerculus marginatus* (Troglodytidae) and a tapaculo *Lioceles thoracicus* (Rhynocriptidae), both share a conspicuous pattern of chestnut plumage with a white bib; a third species, the antbird *Sclerurus albigularis* (Formicariidae), is less strikingly similar. It is also noteworthy that the half-dozen species of toucans (Rhamphastidae) on the Pacific slope of the northern Andes, Colombia, and Panama all have bright yellow bibs, whereas the species of

Amazonia are all white-bibbed. Similarly, three large, flycatching cotingas (Cotingidae) coexist in Peru; although in different genera, they are difficult to distinguish in the field, as all are a uniform grey color. In Panama the three genera are represented by different species which are again found together in forest; again they are difficult to distinguish in the field, but this time all are uniform rufous in color!

Just as the New World hummingbirds Trochilidae are commonly interspecifically territorial (Cody, 1968b), their African counterparts, sunbirds Nectariniidae, exhibit the same phenomenon. Cheke (1971) found that *Nectarinia tacazze* and *N. famosa* have nearly identical diets and defend interspecific territories above 3450 m in Kenya. The two species are alike in voice, and also in shape and proportion. Color differences between the males exist but, as in hummingbirds, do not prevent interspecific aggression.

Character convergence in voice is appropriately discussed next, for on Cerro San Felipe the two towhees *P. ocai* and *P. erythrophthalmus* are also interspecifically territorial, and show similarities in their songs. Each has a number of song types, with a broader range in *ocai,* and most songs are easily assigned to one or the other species. However, just as with the fox sparrow and towhee on Blue Ridge, some song types of each are "intermediate," and cannot be attributed to either except by observing the vocalist. This may have been a recent development, as Charles Sibley (e.g. 1950), who studied *erythrophthalmus-ocai* interrelations extensively over Mexico including Cerro San Felipe, had no such difficulties with vocalist identity. Thus *brunneinucha-ocai* of like appearance and *erythrophthalmus-ocai* of like voice show nonoverlapping territories, even though their habitat requirements where they co-occur are identical. But the pair *brunneinucha-erythrophthalmus* differ in both appearance and voice, and should therefore overlap in territories. They do (Cody and Brown, *op. cit.*). Is there any disadvantage to this overlap? It appears

245

that there is, and that a reduction in available food supply is compensated by an increased territory size over that measured in individuals which by chance do no overlap. Of the five complete *brunneinucha* territories mapped, three did not overlap with *erythrophthalmus* and averaged $(5500 + 5400 + 4400)/3 = 5100$ sq ft, one overlapped partially and measured 8400 sq ft, and the last was completely enclosed by an *erythrophthalmus* territory and was 18,400 sq ft in area.

Character convergence in voice is also known in the wrens of the genus *Thryothorus* and the meadowlarks *Sturnella*. In each case the associated interspecific territoriality has been

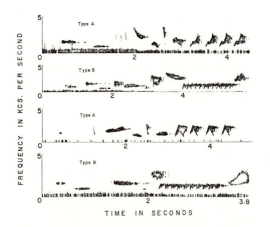

FIGURE 75. Sonograms of wren songs from Mexico. These two species have almost identical songs, and are interspecifically territorial. From Grant, 1966.

documented (Grant, 1966; Lanyon, 1957). *Thryothorus felix* and *T. sinaloa* sing identical songs (Figure 75) and show no territorial overlap on mainland Mexico. Only *T. felix* occurs on the Tres Marias Islands, where it sings a different song, and where its pattern of facial striping resembles that of mainland *sinaloa*. Mainland *T. felix,* however, shows a different facial pattern, indicating that this characteristic might be involved in mate selection, diverging in sympatry and converging in allopatry as the song type does exactly the opposite.

The meadowlarks *Sturnella magna* and *S. neglecta* live in North American grasslands and occupy largely nonoverlapping ranges. Where they meet in the central United States they sing songs intermediate between those heard in allopatric populations where songs are very distinct, and they defend interspecific territories there. Lanyon (1957) found that songs are learned, and thus the proximity of neighbors of different species, if the populations were thoroughly mixed, would account for the intermediate songs. Selection for character convergence here would be tantamount to selection for a) a learned rather than innate song but b) one learned from both congeners. The distribution of the species populations in distinct habitat blocks, which still possess intermediate songs, implies, however, a more extensive selection for an intermediate song per se.

Two other cases of European warblers Sylviidae might be mentioned. Ferry and Deschaintre (1966) found that *Hippolais icterina* and *H. polyglotta,* which are geographical replacements for each other, are interspecifically territorial where they marginally overlap on a northwest-southeast front in eastern France. Their songs there are similar. Likewise Bremond (1970) finds that *Phylloscopus bonelli* and *P. sibilatrix* have very similar songs; these species are extensively allopatric but occur together over large areas in central Europe. A similar situation is described by Thönen (1962) in central Europe among titmice Paridae. The willow tit *Parus montanus* shows a much greater resemblance in song to the marsh tit *Parus palustris,* with which it shares habitat and food, than with its own subspecies the Alpine tit, which lives at higher elevations. The two species also show hostility toward

247

one another. Marsh and willow tits share broad-leafed wood-lands in central Europe, but the former has a more southerly and the latter a more northerly distribution. Of the assemblage of *Parus* in Britain, these two species show the most diet over-lap (75%) of those species which customarily share habitats (Lack, 1971), and so the observed interactions are in accord with our expectations.

3. Song Variation in Passerine Birds. In many passerine spe-cies a great deal of variation in song exists within individuals, within populations, and among populations of the same spe-cies. This sort of variation is particularly common among wrens Troglodytidae, the new world and old world thrushes Mimidae and Turdidae, and the finches Fringillidae and Emberizidae. Marler (1960) recognizes two main functions in bird song—species identity (with attendant implications of territory ownership and readiness to mate) and individual identity. The importance of attaching a personal signature to song was recently demonstrated by Emlen (1971); where neighboring territorial indigo buntings *Passerina* have come to know each other, aggressive encounters between them and the loss of time and energy these imply are minimized. But when strange males sing among them, the aggressive response elicited is much higher. As Marler points out, individuals of a species encounter different sound environments in different parts of their individual, population, and species ranges. They also encounter different sets of competitors. Possibly a partial territorial exclusion among different species through the mech-anism of partial song convergence is widespread in passerine birds, and accounts for some of this song variability. Thus different song types in different habitats or in different parts of the range might be selected to effect a partial exclusion from the territory of different competitors with similar ecologi-cal requirements. Similar reasoning might account for the sort of song mimicry observed in starlings (*Sturnus vulgaris*) and mockingbirds (*Mimus polyglottus*), which incorporate into their song repertoires parts of the songs of other species with which they share the habitat. This might result in the partial exclusion of potential competitors from the territory of the mimic. Clearly in some cases this would not apply, as occasionally the songs of ecologically quite distinct species are picked up and repeated. But those instances might be simply a carry-over from selection for song mimicry in general, and do not negate the possible selective advantages just mentioned.

4. Distributional Patterns in Chilean Birds. The distribution of species over habitats in Chile differs considerably from that of, for instance, Californian birds (Cody, 1970), even though the range of habitats and altitude in the two countries is similar. The most striking difference is the wider use of habitats by Chilean species; even drastic changes of

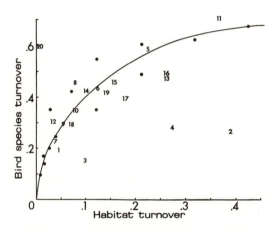

FIGURE 76. The relation between habitat turnover and bird species turnover. Dots and regression line from MacArthur *et al.* 1966. Points 1–4 from Chilean censuses using the same criteria for making comparison, showing very little change in bird species with considerable habitat change. Points 5–20 compare Chilean censuses from very different habitats, altitudes and latitudes, showing that now bird species turnover is comparable to that in North America. From Cody, 1970.

habitat structure (height, density) in the same locality produce little species turnover, i.e., the β-diversity is low (Figure 76). In order to see new species, one must change latitude or, less productively, altitude, as species replacements do occur by geographic area. In contrast, α-diversity is rather high, due at least as much to high species equitability as to high number of species per habitat. Thus the picture is one of habitats well packed with species which persist through habitat changes, a pattern akin in the latter characteristic, but not of course in the former, to what is observed on islands. Central Chile, from where these census data come, is in fact an isolated area. Like islands, it must have a slow colonization rate (by differentiated species in trans-Andean populations), while lacking, because of its large area, the high extinction rates of islands.

In Chile, there are few groups of both ecologically similar and sympatric species, for most congeners are allopatric there. The genera *Muscisaxicola* (10–11 spp.) and *Cinclodes* (6 spp.) and the family Rhinocryptidae (6 spp.) are exceptions, and in these selection appears to have pared down the within-habitat (α-) diversity from a high number of "taxonomic species" to a lower number of "ecological species" by producing interspecific convergences. In this way excess species numbers can, through the medium of behavioral interactions, approach the equilibrium number determined by the habitat structure and its resources. MacArthur and Levins (1967) predict just this phenomenon on theoretical grounds, that, failing a resource space conducive to ecological (and hence morphological) divergence, two species will converge to a common type and utilize resources as a single species. Eventually one or the other species proves competitively superior, and β-diversity is increased if each dominates a particular habitat type.

The phenomenon is illustrated in Figure 77. The Rhinocryptidae are a morphologically homogeneous group in which species size is the main variable. The larger species scratch in leaf litter or excavate around plants, while the smaller

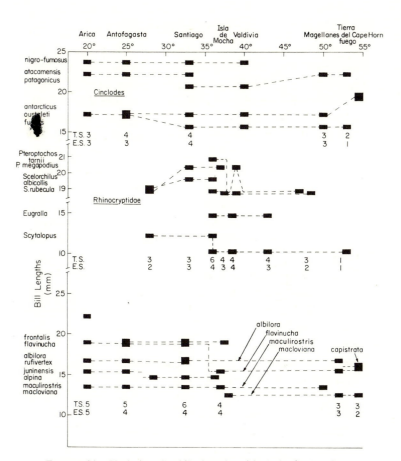

FIGURE 77. Variation in bill length with latitude in three groups of Chilean bird species, the furnariid genus *Cinclodes* (above), dipper-like birds, the family Rhinocryptidae (center), ground feeding birds of thick underbrush, and the tyrannid genus *Muscisaxicola* (below), chat-like birds of heath-tundra. In *Cinclodes*, convergence in bill length occurs between *oustaleti* and *fuscus* in Antofagasta and between *oustaleti* and *antarcticus* on the Cape Horn archipelago. Two species in the largest genera of Rhinocryptidae converge in the north, and another two species in these genera converge in the south. But on competitor-poor Isla de Mocha, the two diverge in bill length. Successive convergences occur in *Muscisaxicola* from north to south such that, by Santiago in central Chile, four species occur in two convergent pairs, whilst two smaller species remain ecologically distinct. These *Muscisaxicola* are known to be interspecifically territorial.

251

may add to these activities the search of low vegetation for insect food. They fill, then, the towhee, thrush, and to some extent the wren niches, and reach maximum diversity in the temperate rain forest of *Nothofagus*. In matorral at Cuesta la Dormida, Santiago province, three species are found, *Scytalopus magellanicus, Scelorchilus albicollis,* and *Pteroptochos megapodius,* with bill sizes 12.3, 19.6, and 20.4 mm, respectively. To the north both larger species change subspecies simultaneously (Atacama), where their measurements are 18.9 and 19.1; they become almost indistinguishable in bill size. In broad-leaf evergreen *Cryptocarya* forest near Concepcion, *Scelorchilus* was not represented, *Pteroptochos* had changed species to *P. tarnii,* and *Scytalopus* was still present. The bill size set becomes 12.3, 15.0, and 20.8 mm, with a new species of medium size, *Eugralla paradoxa.* Further south past the Rio Bio-bio, in *Nothofagus* forest, a *Scelorchilus rubecula* is added with bill size 18.9; but *P. tarnii* changes subspecies to bill size 18.8 and converges with it. Here *Scytalopus* also changes subspecies to parallel the reduction in bill size of the larger species (see Figure 77). In Magellanes only one of the convergent pair, *P. tarnii,* still persists with *Scytalopus,* which latter extends to Tierra del Fuego.

From north to south in the six localities mentioned, it appears that 3, 3, 3, 4, 2, and 1 taxonomic species are compressed by convergences to 2, 3, 3, 3, 2, and 1 ecological species. A result of the convergence in bill morphology might be local competitive exclusion of one of the pair by the other, perhaps mediated by slight habitat differences, thereby reducing α-diversity and increasing β-diversity to more continental norms. Thus on Cerro Ñielol in Temuco, *Scelorchilus* was not found in 1968 (Cody, 1970) and was rare in 1971, although the other three species were all present and common (10.5, 15.0, and 18.8 mm bill sizes). On the other hand, a *Nothofagus* forest near Victoria produced *Scelorchilus* but not *Pteroptochos.*

252

On Mocha Island (lat. 38° 22′) these same four species occur and all are common. But here the convergently similar pairs diverge in size, as the endemic race of *S. rubecula* has bill size 20.4 mm. This divergence is presumably permitted by the reduced competitor diversity there.

The genus *Cinclodes*, ecologically like the dippers Cinclidae but included the South American family Furnariidae, reaches maximum diversity with five species in fresh water and marine habitats near Santiago. By Antofagasta, the middle species (see Figure 77) has dropped out and the smaller two have converged in bill size (from 15.9 and 17.5 to 17.2 and 17.5). Further north only one of the convergent species pair but both of the nonconvergent species still occur. To the south, neither of the larger species reaches Magellanes, and the middle species increases in bill size in apparent compensation for this (21.2 to 22.6) to a bill size intermediate between them. Lastly, of the three species breeding on Tierra del Fuego, only *C. oustaleti* reaches the extreme southern islands and cape, to which the sixth species *C. antarcticus* is restricted (apart from oceanic islands). In the range of *antarcticus* a different subspecies of *oustaleti* is larger (17.5 to 19.7), and matches quite strikingly the 19.5 bill size of *antarcticus*. Thus by convergences taxonomic species numbering 3, 4, 5, 4, 3, 3, 2, respectively, are reduced to 3, 3, 5, 4, 3, 3, and 1.

It is not quite clear how extensive is the range overlap (if any) between the smaller pair of species in Antofagasta, but it is known that the species in the extreme south occur together. The two differ somewhat in appearance and considerably in body size (7″ vs. 9″), but because of the bill size similarity we would expect resource division between the two to be aided by interspecific interactions.

By far the most complex and interesting case of modification of actual species numbers to more appropriate ecological species, numbers, and one in which we have the most extreme field data, exists in the genus *Muscisaxicola*. These are flycatchers of the family Tyrannidae, whose habits strongly resemble those of the chats *Oenanthe* (family Turdidae, see Chapter 5).

Whereas *Cinclodes* reaches a maximum diversity of ecological species in central Chile and the Rhinocryptidae further south, the stronghold of *Muscisaxicola* is in the north. They are species of stony hillsides and valley floors, and all except *M. maculirostris* shun areas with more than a heath-tundra type of vegetation. There is little or no opportunity for habitat segregation other than by altitude. Although at the latitude of Santiago, the six species present can be ranked in the means of their altitudinal distributions, there is great overlap among species, such that usually four species and as many as five species can be found at one place at intermediate altitudes (Cody, 1970). Similarly, at least three species occur together at the same altitude in the north (Smith and Vuilleumier, 1971). The bill size distributions of 12 subspecies of 10 species are shown in Figure 77 for various Chilean latitudes.

In the north of the country in Arica, a remarkably orderly array of bill lengths of the five breeding species exists. But as one moves south to Santiago province, some interesting changes occur. In Antofagasta *juninensis* changes size and subspecies as *alpina,* of a similar size, is added. More startling, when the large-billed *albifrons* is left behind, *frontalis,* which coincides exactly in bill-length with *flavinucha,* is added. By Santiago this process repeats itself, for *albilora* is added; *albilora* differs by only 0.2 mm in bill-length from the medium-sized *rufivertex* (Figure 77).

To put these convergences in perspective, I have plotted in Figure 78 the bill length of the largest member of pairs of adjacent (in ranked size) coexisting species against the bill length of the smaller of the pair. Species pairs from Santiago north to Arica (lumping the convergent species) plus the pair of small species which coexist in the far south, seven points in all, form an orderly relation expected of happily coexisting species with different preferred prey sizes (correlation co-

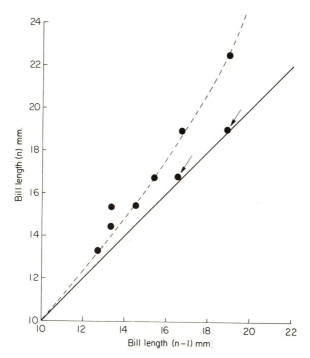

FIGURE 78. Bill lengths of adjacent pairs of species of *Muscisaxicola* in size series in sympatry. Arrows indicate the two convergent pairs of species at Santiago.

efficient 0.98). The ratio of bill lengths l_{x+1}/l_x is not constant, but varies with bill size according to $1.445 - 0.445 \ (11.25/l_x)$, which in the range of bill sizes here varies from 1.05 between the smallest species to 1.22 for the largest species. These ratios are rather smaller than those recorded for temperate passerines by Hutchinson (1959), which averaged 1.28, but larger than those measured by Klopfer and MacArthur (1961) in tropical species pairs. The points representing species pairs *frontalis-flavinucha* and *albilora-rufivertex* lie virtually on the 45° line. Interestingly enough, the species involved in these convergences are those most similar in altitudinal distribution (see Figure 4

255

of Cody, 1970), with the medium-sized pair at medium altitudes and the larger pair at rather higher elevations. Further, in my experience one species in each convergent pair was rare in the five valleys censused; *rufivertex* and *albilora* were respectively the rarest and most common species encountered, while *flavinucha* was second rarest, and *frontalis* was common. Interspecific behavioral interactions were observed between several of the species at intermediate altitudes, and breeding territories were nonoverlapping. Smith and Vuilleumier (1971) also observed interspecific aggression. Although observations were not extensive enough to test this, we would predict that aggressive interactions would be particularly strong and frequent between the pairs of convergent species, and perhaps exist between other pairs of dissimilar bill lengths as a carry-over from selection for the former interactions. All species are uniformly colored brownish-grey, but, due to taxonomic affinities, the application of the term *convergence* to appearance must be withheld. Interspecies recognition for mate selection would surely operate by use of the patches of different colors on the back of the head (present in all species except the two which do not overlap in altitude).

Clearly the five to six taxonomic species found at each latitude between Santiago and Arica are reduced from five to four as one moves from north to south, by successive convergences brought about by species replacements. This reduction is presumed to reflect a more restricted resource space for *Muscisaxicola*, perhaps as simple a matter as increased restriction of the altitudinal range available to the genus. This range is around 16,000' in the north, from near sea level to the altiplano, and around 11,000' in the center of the country (2200' to 13,000', as the snow line is lower and the lower elevations are vegetated).

The situation to the south of Santiago is not nearly so easily resolved. There seems to be a decrease in minimum bill size and a much more marked decrease in maximum bill size in the coexisting species. In Magellanes four species may be found, including a subspecies of *flavinucha* reduced in bill length from 18.9 to 15.2 mm, so that it actually falls far below species it exceeds in the north! Two smaller species, *maculirostris* plus the newcomer *macloviana,* seem to differ by the margin predicted from the Arica species (12.7 versus a predicted 12.5), but the two species of intermediate bill length show no such order (as indicated in Figure 77). There are two possible explanations: these species may occur in separate habitats in the south, or the type of orderly assorting of species observed in the north is still in the process of evolution. This latter seems all the more plausible as colonization of the more recently available habitat in the far south probably occurred both from the north on the western Andean slopes and from the east slope around the bottom of the cordillera.

A case involving similar convergence in morphology has been studied in Costa Rica by Robert Jenkins (pers. comm.). Two species of anis, *Crotophaga sulcirostris* and *C. ani,* now occur together in southwestern Costa Rica, where the latter is pushing north into country formerly occupied solely by the former. Members of the *sulcirostris* population in contact with *ani* are much closer in weight and other measurements of size—total length, wing and leg length, bill size—to *ani* than are individuals further north. Evidence suggests that interspecific territoriality might be involved.

5. *Character Convergence in Other Animal Groups.*

a) Insects. Potential insect candidates for character convergence in interspecific aggression must be territorial or at least spaced, colorful with good visual acuity, and free from confusing Batesian mimicry associations. Dragonflies (Odonata) are thus ideal, and their ecology in North and Central America has been investigated by Dennis Paulson (pers. comm.). A good instance of character convergence appears to involve *Micrathyria eximia* and *Nephepeltia phryne* in Costa Rica,

257

which are identical in the field and appear to be interspecifically territorial. Males of some species treat males of other species as conspecifics, yet distinguish the females. Further work on this sort of behavior is being conducted.

b) Fish. Many coral reef fish, particularly those of the families Chaetodontidae, Pomacentridae, Ballistidae, and Labridae, are territorial. Some of these are extremely brightly colored, but others are not; in some species the young are more brightly colored and more aggressive than the adults. Some species, particularly pomacentrids, have been recorded as interspecifically territorial, but the interpretation of this behavior has been either that it is a misdirected form of intraspecific agonistic behavior (see, e.g., Lorenz, 1966) or that it is simply defense of eggs and nest against potential predators (e.g. Albrech, 1969; Rasa, 1969). *Pomacentrus jenkinsi* was found by Rasa (*op. cit.*) to defend its territory against all intruders, but for us a more significant and interesting behavior is that exhibited by *Pomacentrus flavicauda,* studied on the Great Barrier Reef by Low (1971). *P. flavicauda* excluded 38 species of reef fish from its territory, but ignored another 16 species. The 38 that were excluded varied considerably in appearance, behavior, and taxonomy, but held in common among themselves and with *P. flavicauda* the fact that they are algae-eaters, while the 16 permitted in the territory are all carnivores. Thus the association of the excluded species and their diet similarities with the territory holder is just that described above between interspecifically territorial bird species. This behavior is apparently achieved without convergence in appearance, but in fact many reef species are remarkably alike in their color patterns. It is perhaps significant that the triggerfish observed in an aquarium by Lorenz (1966) reacted aggressively toward two other species, one of which was similar in appearance to the aggressor and the other similar in shape.

c) Amphibians. An apparently clear-cut case of character

convergence occurs between the salamanders *Plethodon jordani* and *Desmognathus ochrophaeus* in the Great Smoky mountains (Cody, 1969). Different subspecies on different mountains possess in common a red "ear" patch and red forelegs, respectively, although not all individuals in the population are so endowed; they are also alike in general body coloration and occur in the same habitat at the same altitudes. Where the two species do not coincide in range or altitude, the incidence of these red markings drops appreciably or altogether. A possible explanation involving Batesian mimicry has received extensive testing but no confirmation. Intraspecific aggression has been observed in both species, and it seems reasonable to attribute their coincident markings to selection for interspecific spacing. Steven Arnold (pers. comm.) finds these species lying with their (distinctively colored) forequarters extended out of their burrows, and rarely more than one individual per log.

The same phenomenon may occur between *Plethodon cinereus* and *D. ochrophaeus* (different subspecies) further north, which possess a red middorsal stripe in common, and between salamander species in the genus *Chiropterotriton* in Mexico. Indeed, it may be widespread. Of particular interest are the many amphibians in which the young are colorful and aggressive and the adults dull and pacific. It appears that, for many amphibians, restricted sites for egg laying and juvenile development are more widely used by different species than are the habitats and feeding places of the adults. In such cases the young of several species may share a simple and undivisible habitat, and may profitably become interspecifically aggressive and spatially assorted.

d) Reptiles. Character convergence has not been reported among reptile species, although because of visual signaling among brightly colored males it might be expected to occur. Kenneth Asplund (pers. comm.) finds that, as with the amphibians mentioned above, the young of sympatric species which later diverge in coloration are initially alike in appearance, and may be more coincident in habitat.

e) Mammals. The only possibility of character convergence among mammals I have encountered (but see Rosenzweig, 1968a) is the resemblance between the ground squirrels *Spermophilus lateralis* and *S. leucurus,* including their close relatives, and the chipmunks *Eutamias.* These are remarkably similar in general coloration, and both have prominently striped sides; lack of facial striping in the ground squirrels distinguishes them from the chipmunks. There are broad areas of sympatry between the two species groups, and at least two chipmunk species and one of the ground squirrel species can be found in the same habitat. Too little is known of their behavioral interactions to discuss the situation further.

D. Social Mimicry

1. The Hypothesis. Through studies of the interspecific social behavior of neotropical birds, Moynihan (1960, 1962, 1963, 1968) came to the conclusion that natural selection may favor within mixed species groups the evolution of signals to promote gregariousness. I use Moynihan's term *social mimicry* for the phenomenon. His explanation of flock formation and cohesion is largely a mechanistic one, and involves the reactions of "circumference" species to individuals or groups of "nucleus" species. The ultimate (as opposed to the proximate) factors involved in flock formation, and in particular flocks of different species, were not discussed, although of course it is a safe assumption that selection has favored flocking in birds and that it is ecologically advantageous. We still need an explanation for the phenomenon that is more than a restatement of faith in this assumption (cf. Lack, 1968, Chap. 4).

2. The Ecological Advantage to Flocking in Birds. Sociality is extremely common among birds. Indeed, the species that do not participate in some sort of flocking behavior at some time during the season or during their lifespan may well be in the minority. A common pattern in many temperate bird species is territorial spacing in the breeding season and flock formation at other times. In tropical forests, flocks may be seen throughout the year; species leave the flocks according to their specific breeding schedules and rejoin them after reproductive efforts cease. Not only do individuals of single species form flocks but in many different situations several species, even dozens, participate in this behavior together. There are potential adaptive advantages of various sorts to flocking, perhaps acting simultaneously, the more subtle of which are discussed next.

Mixed species flocks feeding on small, dispersed food items present the most challenging problem. The mixed species aggregations around clumped, temporarily abundant food supplies mentioned in Section I.A are dismissed, as they form for obvious reasons. Mixed species flocks also require an additional explanation to that offered by Lorenz (1966) for flocks in general, that they facilitate the formation of intraspecific dominance hierarchies and mate selection, which saves economically more important time later when individuals are territorial during the breeding season. More recently the ideas and data of Fretwell (1972) have given impetus to these considerations, but in addition to providing no explanation for mixed species flock formation, they do not account for the fact that individuals already mated and on territories may join flocks for feeding purposes. This occurs in the Mohave Desert, where finch flocks of many species included wintering, migrant, and resident species and continued to attract territory holders in early spring (Cody, 1971). Further, a nonterritorial species, house finch *Carpodacus mexicanus*, travels considerable distances to feed with the flocks.

It is possible that flocks serve some purpose in defense against predators, particularly raptorial birds, by presenting "confusion effects." These can be visual, in which the flock members continually roil around in space (e.g. starlings *Sturnus vulgaris;* Wynne-Edwards, 1962) or simply disperse in all directions, or vocal (e.g. the bushtit *Psaltriparus minimus;* Grinnell, 1903). In either case the predator has increased difficulty in focusing attention on a single individual, a prerequisite for a successful attack.

In the Mohave Desert, initial evidence pointed to feeding ecology as a prime factor in finch flock formation, and further association of differing flock behavior with differing food conditions tends to confirm this (Cody, 1971). The relations are summarized in Figure 51.

Flock formation appears to be a mechanism to regulate the mean "return time" or interval of time between successive visits to a particular point in the habitat. With renewing food supplies, each location in the habitat replenishes its stock after the flock has passed by, and is only worth revisiting after a certain time interval. Flocks can move over the habitat in such a way that the return time mean is regulated to match renewal rate, and its variance minimized. Individuals which forage independently have no knowledge of the history of prior visits to any one feeding site, and thus no way of avoiding short return times except by trial and error.

Three flock variables can affect return times in bounded areas: speed (a negative correlation with return time), the length of straight sections between turns in the path of the flock $(+ve)$, and the deviation from 0.7 of the probability that the flock will continue to go ahead rather than turn; the last result is determined from computer simulation of the flock "walks."

Flocks were studied along a moisture gradient, high at the base of the Granite Mountains (through run-off) and decreasing out into the desert. With increasing distance from these mountains, food abundance decreased and food ripening rate (Cody, 1973b) increased. As the mean return time of the flock should decrease as the ripening rate increases and the food abundance decreases, flocks should return to points in the habitat more rapidly at increasing distance from the mountains. This corresponds with measured flock behavior, as flocks which averaged 540' from the mountains had a return time twice that of flocks which averaged 3560' out. This difference was accomplished by varying all three flock parameters mentioned above, changes related to differences in flock composition in both species and individuals, which in turn is probably governed by the mechanism of differential habitat selection among species.

The general concept of return time regulation by flock formation as a strategy for the exploitation of renewing food supplies may be of more general applicability. The tropical forest flocks of insectivores are of course utilizing renewing insect food, and their paths may show considerable regularity (e.g. Nicholson, 1932). As yet this explanation remains a hypothetical but reasonable interpretation of an otherwise inexplicable phenomena.

3. Social Mimicry in Flocking Birds. Moynihan (1968 and earlier) finds that neotropical bird species which flock together not uncommonly tend to resemble each other in appearance, in spite of the fact that the flock members are of different species from different families. Thus lowland forest flocks in Panama, which comprise antbirds (family Formicariidae), a tanager (Thraupidae), and a warbler (Parulidae), show considerable uniformity. Montane bush flocks in western Panama contain a preponderance of species with black and/or yellow markings, which include finches, tanagers, and warblers. In the humid temperate forest of the northern Andes the members of mixed-species flocks are predominantly either blue or blue and yellow (tanagers, and honeycreepers Coeribidae), while in the same habitat further south species are commonly blue or blue-grey above with chestnut or buff underparts (tanagers, honeycreepers, and a plush-capped finch Catamblyrhynchidae). Another instance of such convergence on a grand scale is reported by Terborgh (pers. comm.) from lowland rain forest in Peru. Of 26–30 species of antbirds (Formicariidae), 11, in four different genera, are understory foragers that frequently join mixed-species flocks. The males of eight of these species are uniformly dull blue-grey in color, the other three are charcoal; they are distinguished only by embellishment by small black and white flecks. The similarities among species are far too striking to be mere coincidence, and an argument invoking common selection for crypticity is rendered unlikely by the fact that the females of all 11 species are similarly reddish and greyish-browns, but much more readily distinguished than the males. The semi-social coots (*Fulica*), which are intra- and interspecifically aggressive at short range (Cody, 1970) and show little or no divergence in appearance, may be a further example.

Moynihan also believes that the association of common color pattern with membership in a mixed-species flock can hardly be coincidental, and it is likely that selection has favored convergence to promote the aggregations.

The appearances of the species that occur in the Mohave Desert finch flocks are likewise similar, but this can hardly be called social mimicry. For one thing, all the species are related to the extent that they all belong to the same family *Emberizidae*, and further, the coloration of them all is predominantly the dull browns and greys typical of almost all North American sparrows. The call notes of the participant species, however, do show similarities, and may have converged under natural selection; the flocks maintain their integration and cohesion chiefly through the use of such vocalizations.

4. Convergence to Facilitate Both Aggression and Aggregation. One of the most striking cases of social mimicry is that of the Panamanian grassland finches (Moynihan, 1960), in which the characteristics of all black plumage with a white flash on the wing are retained across three genera, *Volatinia*, *Sporophila*, and *Oryzoborus*. These form mixed-species flocks, and at least some species in some combinations are interspecifically territorial (N. G. Smith, pers. comm.). Thus the convergent appearance might serve equally well to separate breeding individuals as it would to promote flocking outside the breeding season. The tyrant flycatchers *Muscisaxicola* are a second example of the same phenomenon. As described above, all species are similar in appearance and are distinguished chiefly by a dab of color at the back of the head. Not only are they interspecifically territorial during the breeding season but in the months preceding the establishment of territories they form mixed-species flocks in the Andean foothills (Cody, 1970). These associations were also noted by Smith and Vuilleumier (1971). Again, the similarity in appearance could function equally well in promoting aggression between species at one season as aggregation at another, but, as the species are so closely related, it is more appropriate to speak of lack of divergence than of convergence.

III. CONCLUSION

Species that converge in appearance, voice, and/or morphology are apparently responding to a resource space unable to support them as separate ecological entities. This is particularly apparent in the case of the *Muscisaxicola* species, in which the number of "ecological species" matches the altitudinal range available to the genus better than the number of taxonomic species. Likewise, the towhee niche in the San Gabriel Mountains supports two species at lower elevations that apparently coexist without interaction. At higher elevations one species drops out, and at still higher elevations two species occupy the niche but are convergently similar in song and partially interspecifically territorial. The phenomenon of interspecific territoriality is particularly common where two species marginally overlap in ranges which elsewhere support just a single species.

Flock formation is likewise associated with a shortage in food resources, both of variety and of abundance. The Mohave Desert finches all eat the same seed foods, and flocks become larger and develop from single species to mixed-species aggregations as the winter season progresses and the nonrenewing food supplies become scarcer. Groups of species in the flocks behave ecologically as single species, and convergence to a common mode of food acquisition is again associated with food resource shortage. The switch from a territorial to a flocking system commonly observed in many North American bird species is generally associated with a change in food supplies from the season in which they are most abundant to that in which they are most severely taxed. The same may be true of the neotropical flocks which show social mimicry, although it remains to be shown that displacement patterns among those species are not viable alternatives eliminated through selection for the most efficient method of insect harvesting.

At the other end of the resource abundance spectrum in situations of temporarily superabundant food supplies, different species may show similarities in the ways in which these food supplies are exploited. Although there may be no actual convergence involved in the sense discussed above, there is certainly no selection for divergence, and once more the displacement patterns discussed earlier in this book and which are the dominant and expected patterns in species assemblages fail to emerge. Displacement patterns, the result of selection for divergence in ecological characteristics particularly associated with the ways in which food is obtained, evolve only on resources which are neither superabundant nor extremely low, either in predictability or in abundance. Toward either extreme in the resource state, selection no longer favors divergence; on superabundant resources neither divergence nor convergence is selected, but rather species exploit the resources opportunistically. When resources are in short supply relative to the numbers of individuals or species exploiting them, convergence results both within and between species to flocking in the nonbreeding season and between species to interspecific territoriality in the breeding season.

APPENDIX A

Scrub Community Data

1. Wyoming Willows

a) Site: Jackson Hole, Grand Teton National Park, Wyoming. Lat.43°52′N, Long.110°34′W. Elevation 6790′. Area 5.6 acres.

b) Vegetation: Many *Salix* species. Clearings with forbs (e.g. *Castilleja, Aconitum, Solidago*), sedges, and grasses.

c) Climate: a) Year-to-year variance in breeding rainfall $R_V = 0.504$; b) Year-to-year variance in breeding season temperature $T_V = 5.73$; c) Probability of encountering aseasonal rainfall or temperature $P = 0.094$; d) Rainfall seasonality $R(M - M) = 2.42$; e) Temperature seasonality $T(M - M) = 5.03$; f) Length of growing season, from average number of frost-free days/year $FFD = 35$.

	Body size[1]
d) Bird species:	
1. Calliope hummingbird *Stellula calliope*	34.7 mm
2. Wilson's warbler *Wilsonia pusilla*	50.6
3. Yellowthroat *Geothlypis trichas*	53.7
4. Yellow warbler *Dendroica petechia*	57.0
5. Clay-colored sparrow *Spizella pallida*	59.0
6. Trail's flycatcher *Empidonax trailii*	61.8
7. Lincoln's sparrow *Melospiza lincolnii*	65.0
8. Song sparrow *Melospiza melodia*	69.1
9. Cliff swallow *Petrochelidon pyrrhonota*	70.9

[1] Body size: total length minus tail length minus bill length.

10. White-crowned sparrow *Zonotrichia leucophrys* — 74.4
11. Fox sparrow *Passerella iliaca* — 76.4
12. Swainson's thrush *Hylocichla ustulata* — 82.9

2. *Wyoming Sagebrush*

a) Site: Jackson Hole, Grand Teton National Park, Wyoming Lat.43°52'N. Long.110°34'W. Elevation 6795'. Area 12.0 acres.
b) Vegetation: Mostly *Artemesia* species. Various grasses.
c) Climate: as above for Wyoming Willows.
d) Bird species:

1. Brewer's sparrow *Spizella breweri* — 55.2
2. Cliff swallow *Petrochelidon pyrrhonota* — 70.9
3. Vesper sparrow *Poocetes gramineus* — 71.8
4. White-crowned sparrow *Zonotrichia leucophrys* — 74.4
5. Brewer's blackbird *Euphagus cyanocephalus* — 107.2

3. *Colorado Saltbush*

a) Site: Pawnee National Grasslands, Weld County, northeastern Colorado. Lat.40°50'N, Long.104°45'W. Elevation 4700'. Area 10.1 acres.
b) Vegetation: *Atriplex canescens*, occasional *Artemesia;* various grasses (*Bouteloua, Buchloe, Agropyron, Stipa*).
c) Climate: a) $R_V = 2.036$; b) $T_V = 8.41$; c) $P = 0.106$; d) $R(M - M) = 6.96$; e) $T(M - M) = 3.52$; f) $FFD = 125$.
d) Bird species:

1. Brewer's sparrow *Spizella breweri* — 55.2
2. Horned lark *Eremophila alpestris* — 78.0
3. Lark bunting *Calamospiza melanocorys* — 83.7
4. Sage thrasher *Oreoscoptes montanus* — 90.8
5. Western meadowlark *Sturnella neglecta* — 123.7

4. *Mohave Desert*

a) Site: 25 miles north of Amboy, SE California, at foot of Granite Mountains. Lat.34°47′N, Long.115°39′W. Elevation 4240′. Area 9.0 acres.

b) Vegetation: Mixed desert shrubs, including *Larrea, Acacia greggii, Yucca mohavensis, Ephedra* spp., *Happlopappus* spp., *Coleogyne, Eriogonum* spp., *Tetradymia*.

c) Climate: a) $R_V = 0.153$; b) $T_V = 7.36$; c) $P = 0.112$; d) $R(M - M) = 4.64$; e) $T(M - M) = 2.27$; f) $FFD = 238$.

d) Bird species:

1.	Costa's hummingbird *Calypte costae*	43.2
2.	Blue-gray gnatcatcher *Polioptila caerulea*	46.0
3.	Verdin *Auriparus flaviceps*	50.3
4.	Black-throated sparrow *Amphispiza bilineata*	56.1
5.	Violet-green swallow *Tachycineta thalassina*	64.9
6.	House finch *Carpodacus mexicanus*	70.9
7.	Phainopepla *Phainopepla nitens*	77.5
8.	White-throated swift *Aeronautes saxatalis*	83.6
9.	Loggerhead shrike *Lanius ludovicianus*	92.9
10.	LeConte's thrasher *Toxostoma lecontei*	100.5
11.	Mourning dove *Zenaida macroura*	145.9
12.	Gambel's quail *Lophortyx gambelii*	147.3

5. *California Chaparral*

a) Site: Murphy Ranch, Santa Monica Mountains, 5 miles south of Calabasas, S California. Lat.34°03′N, Long.118°39′W. Elevation 2075′, Area 7.4 acres.

b) Vegetation: Mixed needle-leaf and broad-leaf shrubs, evergreen and sclerophyllous. Including *Ceanothus* spp.,

Rhus spp., *Arctostaphyllos* spp., *Quercus dumosa, Adeno-stoma* spp. and *Heteromeles arbutifolia.*

c) Climate: a) $R_V = 0.099$; b) $T_V = 3.55$; c) $P = 0.028$;
   d) $R(M - M) = 11.19$; e) $T(M - M) = 0.78$; f)
   $FFD = 365$.

d) Bird species:

1. Bushtit *Psaltriparus minimus* 43.8
2. Anna's hummingbird *Calypte anna* 46.5
3. Bewick's wren *Thryomanes bewickii* 51.5
4. Orange-crowned warbler *Vermivora celata* 53.3
5. Wrentit *Chamaea fasciata* 57.3
6. Plain titmouse *Parus inornatus* 58.9
7. Solitary vireo *Vireo solitarius* 60.7
8. White-throated swift *Aeronautes saxatalis* 85.4
9. Rufous-sided (spotted) towhee *Pipilo erythrophthalmus* 85.4
10. Ash-throated flycatcher *Myiarchus cinerascens* 85.4
11. Black-headed grosbeak *Pheuticus melanocephalus* 86.3
12. Brown towhee *Pipilo fuscus* 86.6
13. Nuttall's woodpecker *Dendrocopus nuttallii* 88.4
14. California thrasher *Toxostoma redivivum* 118.0
15. Scrub jay *Aphelocoma coerulescens* 119.7
16. Red-shafted flicker *Colaptes cafer* 129.5
17. California quail *Lophortyx californicus* 131.4

6. Lower Sonoran Desert
a) Site: 11 miles NE of Tucson, Arizona. Lat.32°15′N, Long.110°46′W. Elevation 2330′. Area 13.7 acres.

b) Vegetation: Many cacti (Saguaro, *Cereus giganteus;* cholla, *Opuntia* spp.), some trees *(Cercidium, Prosopis)* many shrubs *(Larrea, Lycium, Gutierrezia).*

c) Climate: a) $R_V = 0.541$; b) $T_V = 5.44$; c) $P = 0.033$; d) $R(M - M) = 6.40$; e) $T(M - M) = 2.06$; f) $FFD = 229$.

d) Bird species:

1. Black-chinned hummingbird *Archilochus alexandri*	41.2
2. Black-tailed gnatcatcher *Polioptila melanura*	43.2
3. Lucy's warbler *Vermivora luciae*	47.0
4. Verdin *Auriparus flaviceps*	50.3
5. House finch *Carpodacus mexicanus*	70.9
6. Pyrrhuloxia *Pyrrhuloxia sinuata*	82.0
7. Ash-throated flycatcher *Myiarchus cinerascens*	85.4
8. Cactus wren *Campylorhynchus brunneicapillus*	92.2
9. Purple martin *Progne subis*	100.1
10. Lesser nighthawk *Chordeiles acutipennis*	105.5
11. Gila woodpecker *Centurus uropygialis*	113.7
12. Curve-billed thrasher *Toxostoma curvirostre*	115.6
13. Gilded flicker *Colaptes chrysoides*	137.4
14. Mourning dove *Zenaidura macroura*	145.9
15. Gambel's quail *Lophortyx gambelii*	147.3
16. White-winged dove *Zenaida asiatica*	148.8

7. *Arizona Mesquite*

a) Site: 1 mile E of Portal, SE Arizona. Lat.31°58′N, Long.-109°07′W. Elevation 4800′. Area 13.8 acres.

b) Vegetation: Mesquite *Prosopis juliflora, Acacia greggii* and *A. constricta, Ephedra, Atriplex, Sapindus.* Grasses *Munroa* and *Setaria.*

c) Climate: a) $R_V = 0.868$; b) $T_V = 5.34$; c) $P = 0.038$; d) $R(M - M) = 6.31$; e) $T(M - M) = 2.30$; f) $FFD = 231$.

d) Bird species:

1. Black-chinned hummingbird
   Archilochus alexandri   41.2
2. Black-tailed gnatcatcher *Polioptila*
   melanura   43.1
3. Black-throated sparrow *Amphispiza*
   bilineata   56.1
4. Pyrrhuloxia *Pyrrhuloxia sinuata*   82.0
5. White-throated swift *Aeronautes*
   saxatalis   83.6
6. Ash-throated flycatcher *Myiarchus*
   cinerascens   85.4
7. Ladder-backed woodpecker
   Dendrocopus scalaris   86.5
8. Brown towhee *Pipilo fuscus*   86.6
9. Cactus wren *Campylorhynchus*
   brunneicapillus   92.2
10. Bendire's thrasher *Toxostoma bendirei*   104.3
11. Mockingbird *Mimus polyglotta*   105.0
12. Lesser nighthawk *Chordeiles*
    acutipennis   105.5
13. Mourning dove *Zenaidura macroura*   145.9
14. Gambel's quail *Lophortyx gambelii*   147.3

8. Arizona Pine-Oak

a) Site: ¼ mile east of the Southwestern Research Station, Chiricahua Mountains, SE Arizona. Lat.31°51′N, Long.109°10′W. Elevation 5400′. Area 6.75 acres.

b) Vegetation: Pines (*Pinus leiophylla*, *P. engelmanni*), oaks (*Quercus arizonica*, *Q. emoryi*, *Q. hypoleucoides*), juniper (*Juniperus deppeana*).

c) Climate: a) $R_V = 1.465$; b) $T_V = 3.01$; c) $P = 0.013$;

d) $R(M - M) = 7.87$; e) $T(M - M) = 2.37$; f) $FFD = 154$.

d) Bird species:

1.	Black-chinned hummingbird *Archilochus alexandri*	41.2
2.	Pygmy nuthatch *Sitta pygmaeus*	43.1
3.	Bushtit *Psaltriparus minimus*	43.6
4.	Bridled titmouse *Parus wollweberi*	50.0
5.	Black-throated gray warbler *Dendroica nigrescens*	52.8
6.	Painted redstart *Setophaga picta*	52.8
7.	Bewick's wren *Thryothorus bewickii*	55.2
8.	Grace's warbler *Dendroica graciae*	55.3
9.	Solitary vireo *Vireo solitarius*	61.1
10.	Chipping sparrow *Spizella passerina*	62.1
11.	White-breasted nuthatch *Sitta carolinensis*	63.5
12.	Western wood peewee *Contopus sordidulus*	64.1
13.	Violet-green swallow *Tachycineta thalassina*	64.9
14.	Olivaceous flycatcher *Myiarchus tuberculifer*	65.9
15.	Eastern bluebird *Sialia sialis*	79.0
16.	White-throated swift *Aeronautes saxatalis*	83.6
17.	Hepatic tanager *Piranga flava*	87.2
18.	Acorn woodpecker *Melanerpes formicivorus*	107.1
19.	Mexican jay *Aphelocoma ultramarina*	124.0
20.	Red-shafted flicker *Colaptes cafer*	129.5

9. Coastal Scrub, Chile

a) Site: 2 miles east of Pichidangui, Coquimbo Province, central Chile. Lat.32°09'S, Long.71°33'W. Elevation 60', Area 7.5 acres.

b) Vegetation: Shrubs of sagebrush size: *Bahia, Happlopappus* sp.
c) Climate: Data are incomplete on all three Chilean sites.
d) Bird species:

   1. Band-tailed sierra finch *Phrygilus alaudinus*[2]   85.1
   2. Diuca finch *Diuca diuca*   92.4
   3. Red-breasted meadowlark *Pezites militaris*   147.7
   4. Chilean mockingbird *Mimus tenca*   154.5
   5. Great shrike-tyrant *Agriornis livida*   179.4

10. Matorral, Chile
a) Site: 4 miles east of Puchuncavi, Valparaiso Province, central Chile. Lat.32°45′S, Long.71°23′W. Elevation 1050′. Area 4.6 acres.
b) Vegetation: Mixed needle-leaf and broad-leaf shrubs. Evergreen and sclerophyllous. Including *Kageneckia, Quillaja, Trevoa, Lithraea, Colliguaya, Satureja.*
c) Climate: No details available.
d) Bird species:

   1. Tufted tit-tyrant *Anaeretes parulus*   58.1
   2. House wren *Troglodytes aedon*   63.3
   3. Plain-mantled tit-spinetail *Leptasthenura aegithaloides*   65.7
   4. White-crested elaenia *Elaenia albiceps*   78.1
   5. Chilean swallow *Tachycineta leucopyga*   81.0
   6. Rufous-collared sparrow *Zonotrichia capensis*   81.5
   7. Giant hummingbird *Patagona gigas*   84.2
   8. Diuca finch *Diuca diuca*   92.4

[2] Names follow R. M. DeSchauensee, 1970, *A Field Guide to the Birds of South America.*

9. Rufous-tailed plantcutter *Phytotoma
   rara*   101.5
10. Dusky-tailed canastero *Asthenes
    humicola*   104.3
11. Band-winged nightjar *Caprimulgus
    longirostris*   116.3
12. Fire-eyed diucon *Pyrope pyrope*   117.2
13. Moustached turca *Pteroptochos
    megapodius*   145.0
14. Austral blackbird *Curaeus curaeus*   148.0
15. Chilean mockingbird *Mimus tenca*   154.5
16. Eared dove *Zenaidura auriculata*   154.5
17. California quail *Lophortyx californicus* 157.2
18. Austral thrush *Turdus falklandii*   172.5
19. Great shrike-tyrant *Agriornis livida*   179.4

11. Chilean Savannah

a) Site: 8 miles south of Melipilla, Santiago Province, central
   Chile. Lat.33°40'S, Long.71°15'W. Elevation 520'.
   Area 16.4 acres.
b) Vegetation: *Prosopis* and *Acacia caven*. This habitat is possibly degenerate matorral.
c) Climate: No details available.
d) Bird species:

   1. Tufted tit-tyrant *Anaeretes parulus*   58.1
   2. House wren *Troglodytes aedon*   63.3
   3. Plain-mantled tit-spinetail
      Leptasthenura aegithaloides   65.7
   4. Andean tapaculo *Scytalopus
      magellanicus*   66.5
   5. White-crested elaenia *Elaenia albiceps* 78.1
   6. Rufous-collared sparrow *Zonotrichia
      capensis*   81.5
   7. Diuca finch *Diuca diuca*   92.4
   8. Dusky-tailed canastero *Asthenes
      humicola*   104.3

APPENDIX B

Community Matrices And Dendrograms

1. WYOMING WILLOWS

Community matrix and dendrogram in text, Fig.32.

2. WYOMING SAGEBRUSH

	1	2	3	4	5
1	1.000	0.387	0.685	0.760	0.278
2		1.000	0.386	0.282	0.305
3			1.000	0.611	0.359
4				1.000	0.258
5					1.000

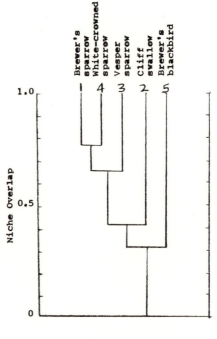

3. <u>COLORADO</u> <u>SALTBUSH</u>

	1	2	3	4	5
1	1.000	0.456	0.388	0.418	0.244
2		1.000	0.644	0.572	0.559
3			1.000	0.456	0.552
4				1.000	0.419
5					1.000

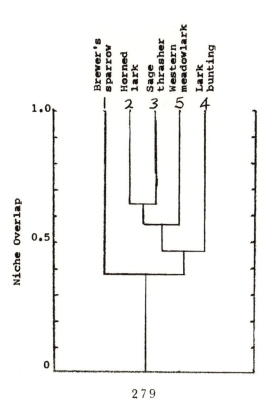

APPENDIX B

4. <u>MOHAVE DESERT</u>

	1	2	3	4	5	6	7	8	9	10	11	12
1	1.000	0.658	0.637	0.482	0.366	0.416	0.579	0.389	0.412	0.416	0.372	0.497
2		1.000	0.610	0.472	0.364	0.430	0.511	0.386	0.474	0.337	0.404	0.446
3			1.000	0.524	0.344	0.448	0.727	0.357	0.380	0.217	0.423	0.425
4				1.000	0.454	0.867	0.411	0.446	0.791	0.594	0.703	0.864
5					1.000	0.419	0.268	0.537	0.371	0.330	0.308	0.483
6						1.000	0.396	0.411	0.718	0.551	0.626	0.759
7							1.000	0.278	0.346	0.277	0.476	0.337
8								1.000	0.365	0.340	0.330	0.539
9									1.000	0.576	0.585	0.777
10										1.000	0.484	0.635
11											1.000	0.765
12												1.000

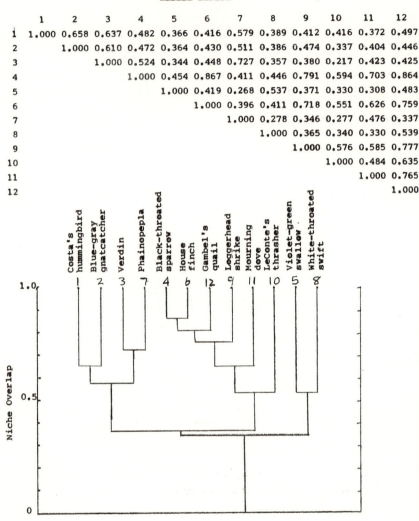

5. CALIFORNIA CHAPARRAL

Community dendrogram in text, Fig.59.

	4	5	6	7	8	9	10	11	12	13	14	15	16	17
1	0.569	0.768	0.667	0.527	0.456	0.479	0.643	0.588	0.456	0.517	0.396	0.505	0.397	0.497
2	0.352	0.656	0.434	0.267	0.402	0.398	0.487	0.449	0.399	0.463	0.418	0.484	0.401	0.451
3	0.409	0.788	0.507	0.345	0.414	0.480	0.568	0.558	0.449	0.492	0.451	0.560	0.466	0.545
4	1.000	0.486	0.782	0.705	0.265	0.412	0.472	0.390	0.386	0.590	0.245	0.384	0.257	0.350
5		1.000	0.590	0.441	0.507	0.555	0.573	0.638	0.539	0.482	0.474	0.606	0.482	0.589
6			1.000	0.635	0.327	0.492	0.539	0.443	0.470	0.662	0.307	0.468	0.307	0.438
7				1.000	0.232	0.349	0.331	0.367	0.296	0.581	0.215	0.290	0.243	0.328
8					1.000	0.382	0.301	0.281	0.350	0.302	0.358	0.375	0.327	0.433
9						1.000	0.339	0.507	0.794	0.350	0.704	0.633	0.488	0.645
10							1.000	0.580	0.346	0.407	0.344	0.474	0.370	0.313
11								1.000	0.492	0.551	0.411	0.666	0.437	0.430
12									1.000	0.431	0.625	0.688	0.570	0.683
13										1.000	0.398	0.555	0.394	0.372
14											1.000	0.672	0.671	0.534
15												1.000	0.767	0.733
16													1.000	0.670
17														1.000

	1	2	3
1	1.000	0.539	0.677
2		1.000	0.716
3			1.000

6. SONORAN DESERT

	1	2	3	4	5	6	7	8	9	10	11	12	13	14	15	16
1	1.000	0.489	0.318	0.503	0.444	0.506	0.657	0.568	0.461	0.386	0.637	0.466	0.453	0.384	0.325	0.450
2		1.000	0.495	0.675	0.635	0.572	0.640	0.530	0.366	0.483	0.542	0.461	0.443	0.475	0.455	0.411
3			1.000	0.615	0.335	0.324	0.399	0.429	0.318	0.417	0.483	0.418	0.343	0.456	0.337	0.287
4				1.000	0.623	0.610	0.672	0.707	0.475	0.658	0.622	0.555	0.512	0.488	0.504	0.472
5					1.000	0.623	0.602	0.678	0.485	0.514	0.533	0.520	0.507	0.556	0.591	0.548
6						1.000	0.594	0.553	0.345	0.505	0.491	0.507	0.329	0.429	0.456	0.374
7							1.000	0.663	0.363	0.554	0.659	0.387	0.496	0.411	0.383	0.487
8								1.000	0.430	0.548	0.717	0.524	0.576	0.579	0.582	0.677
9									1.000	0.684	0.433	0.762	0.380	0.543	0.496	0.545
10										1.000	0.574	0.331	0.460	0.493	0.481	0.431
11											1.000	0.693	0.434	0.507	0.439	0.579
12												1.000	0.665	0.609	0.608	0.647
13													1.000	0.810	0.677	0.629
14														1.000	0.765	0.565
15															1.000	0.596
16																1.000

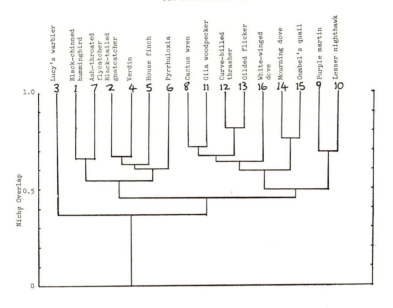

7. ARIZONA MESQUITE

	3	4	5	6	7	8	9	10	11	12	13	14	
1	0.366	0.345	0.281	0.487	0.366	0.330	0.367	0.447	0.521	0.438	0.290	0.149	1
2	0.328	0.494	0.277	0.419	0.492	0.267	0.337	0.181	0.603	0.335	0.439	0.227	2
3	1.000	0.403	0.548	0.529	0.480	0.837	0.622	0.537	0.508	0.489	0.662	0.545	3
4		1.000	0.279	0.524	0.518	0.429	0.467	0.231	0.536	0.373	0.368	0.243	4
5			1.000	0.340	0.368	0.405	0.348	0.330	0.430	0.585	0.446	0.292	5
6				1.000	0.603	0.491	0.615	0.377	0.642	0.469	0.433	0.324	6
7					1.000	0.477	0.622	0.361	0.776	0.437	0.415	0.309	7
8						1.000	0.600	0.564	0.504	0.415	0.669	0.570	8
9							1.000	0.416	0.568	0.391	0.493	0.381	9
10								1.000	0.433	0.404	0.696	0.607	10
11									1.000	0.517	0.509	0.347	11
12										1.000	0.441	0.291	12
13											1.000	0.750	13
14												1.000	14

	1	2
1	1.000	0.385
2		1.000

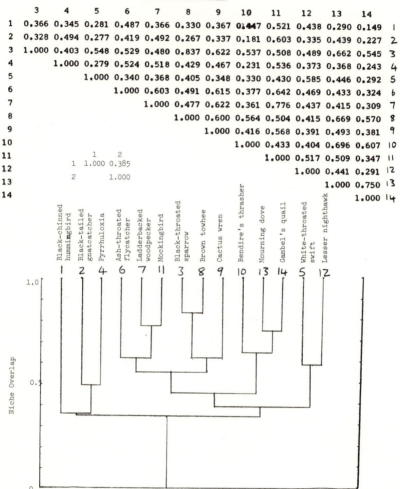

Niche Overlap

Black-chinned hummingbird 1 · Black-tailed gnatcatcher 2 · Pyrrhuloxia 4 · Ash-throated flycatcher 6 · Ladderbacked woodpecker 7 · Mockingbird 11 · Black-throated sparrow 3 · Brown towhee 8 · Cactus wren 9 · Bendire's thrasher 10 · Mourning dove 13 · Gambel's quail 14 · White-throated swift 5 · Lesser nighthawk 12

APPENDIX B

8. <u>ARIZONA PINE-OAK</u>

Community matrix and dendrogram in text, Fig.33.

9. <u>COASTAL SCRUB, CHILE</u>

	1	2	3	4	5
1	1.000	0.598	0.273	0.434	0.281
2		1.000	0.469	0.451	0.314
3			1.000	0.486	0.410
4				1.000	0.762
5					1.000

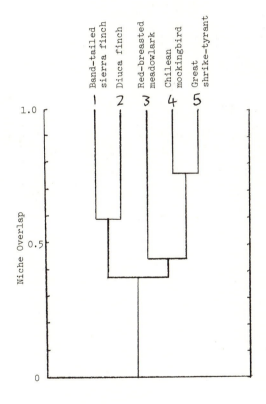

10. <u>MATORRAL</u>, <u>CHILE</u>

Community dendrogram in text, Fig.60.

	6	7	8	9	10	11	12	13	14	15	16	17	18	19
1	0.645	0.459	0.457	0.654	0.720	0.453	0.629	0.504	0.366	0.663	0.495	0.548	0.537	0.345
2	0.692	0.473	0.533	0.589	0.759	0.656	0.592	0.466	0.419	0.596	0.487	0.564	0.591	0.365
3	0.580	0.536	0.416	0.568	0.626	0.395	0.639	0.423	0.321	0.529	0.396	0.468	0.480	0.265
4	0.548	0.389	0.394	0.699	0.660	0.391	0.675	0.457	0.346	0.630	0.396	0.483	0.452	0.305
5	0.451	0.338	0.348	0.343	0.364	0.798	0.375	0.414	0.286	0.378	0.427	0.471	0.392	0.268
6	1.000	0.367	0.736	0.593	0.758	0.453	0.528	0.580	0.480	0.682	0.585	0.749	0.695	0.424
7		1.000	0.257	0.438	0.409	0.385	0.489	0.370	0.326	0.398	0.387	0.352	0.393	0.279
8			1.000	0.389	0.570	0.356	0.345	0.520	0.681	0.558	0.653	0.708	0.663	0.624
9				1.000	0.630	0.383	0.560	0.419	0.252	0.538	0.435	0.437	0.444	0.238
10					1.000	0.387	0.551	0.548	0.457	0.742	0.595	0.590	0.633	0.397
11						1.000	0.417	0.422	0.298	0.391	0.447	0.482	0.406	0.260
12							1.000	0.514	0.315	0.533	0.377	0.457	0.458	0.359
13								1.000	0.558	0.718	0.580	0.668	0.707	0.558
14									1.000	0.558	0.621	0.548	0.582	0.759
15										1.000	0.649	0.635	0.718	0.500
16											1.000	0.671	0.705	0.564
17												1.000	0.736	0.482
18													1.000	0.485
19														1.000

	1	2	3	4	5
1	1.000	0.754	0.709	0.816	0.505
2		1.000	0.753	0.713	0.389
3			1.000	0.661	0.354
4				1.000	0.358
5					1.000

11. SAVANNAH, CHILE

	1	2	3	4	5	6	7	8	9	10	11	12	13	14	15	16
1	1.000	0.522	0.676	0.304	0.532	0.528	0.515	0.574	0.479	0.497	0.381	0.326	0.498	0.455	0.423	0.251
2		1.000	0.612	0.249	0.513	0.504	0.457	0.513	0.449	0.334	0.342	0.303	0.357	0.343	0.304	0.122
3			1.000	0.340	0.653	0.533	0.492	0.538	0.638	0.480	0.344	0.357	0.467	0.370	0.492	0.171
4				1.000	0.179	0.385	0.440	0.433	0.272	0.280	0.367	0.446	0.493	0.647	0.524	0.430
5					1.000	0.453	0.423	0.572	0.478	0.538	0.294	0.198	0.410	0.282	0.320	0.244
6						1.000	0.860	0.737	0.458	0.491	0.656	0.610	0.655	0.619	0.557	0.394
7							1.000	0.663	0.475	0.483	0.679	0.624	0.634	0.650	0.697	0.397
8								1.000	0.561	0.617	0.561	0.476	0.657	0.585	0.535	0.459
9									1.000	0.535	0.434	0.415	0.456	0.424	0.383	0.237
10										1.000	0.355	0.335	0.479	0.469	0.453	0.401
11											1.000	0.911	0.761	0.628	0.588	0.459
12												1.000	0.731	0.629	0.425	0.681
13													1.000	0.631	0.488	0.577
14														1.000	0.586	0.348
15															1.000	0.348
16																1.000

APPENDIX B

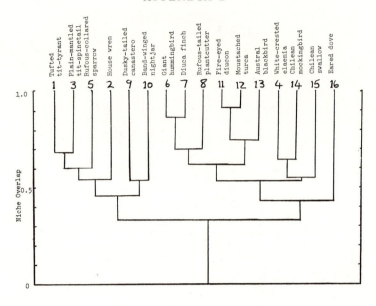

References

Albrecht, H., 1969. Behavior of four species of Atlantic damselfishes from Colombia, South America (*Abudefduf saxatiles, A. taurus, Chromis multilineata, C. cyanea;* Pisces, Pomacentridae). *Z. Tierpsychol.* 26:662–676.

Alm, B., H. Myhrberg, E. Nyholm, and S. Svenson, 1966. Densities of birds in Alpine heaths. *Vår Fågelv.* 25:193–201.

Ashmole, N. P., and M. J. Ashmole, 1967. Comparative feeding ecology of sea birds of a tropical oceanic island. *Peabody Mus. Nat. Hist. Bull.* 24:1–131.

Baird, S. F., T. M. Brewer, and R. Ridgeway, 1874. *A History of North American Birds.* Vol. 1, Land Birds.

Bédard, J., 1969a. Feeding of the least, crested and parakeet auklets around St. Lawrence Island, Alaska. *Canad. J. Zool.* 47:1025–1050.

Bédard, J., 1969b. Adaptive radiation of the Alcidae. Ibis 111:189–198.

Bédard, J., 1969c. Histoire naturelle du Gode, *Alca torda,* dans le golfe Sainte Laurant, province de Quebec, Canada. *Etude du service canadien de la faune,* No. 7. Ottawa.

Belopolsk'ii, L. O., 1957, 1961. *The Ecology of sea-colony birds of the Barents Sea.* Israel Program for Scient. Trans. 1–136.

Bernard, F., 1961. Biotopes habituels des fourmis sahariennes de plaine, d'apres l'abondance de leurs nids en 60 stations tres diversés. *Bull. Soc. d'Hist. Nat. de l'Afrique du Nord* 52:21–40.

Bernstein, R. A., 1971. *The ecology of ants in the Mohave Desert: Their interspecific relationships, resource utilization and diversity.* Ph.D. Thesis, University of Calif., Los Angeles.

Blondel, J., 1965. Étude des populations d'oiseaux dans une garrigue Méditerranéanne: Desciption du milieu, de la methode de travail et expose des premiers resultats obtenus a la periode de reproduction. *Terre et Vie* 112:311–341.

Blondel, J., 1969. *Synecologie des Passereaux Residents et Migrateurs dans Le Midi Méditerranéen Francais.* Centre Regional de Documentation Pedagogique. Marseille.

Bowman, R. I., 1963. Evolutionary patterns in Darwin's finches. *Occ. Papers Calif. Acad. Sci.* 44:107–140.

Brémond, J. C., 1970. Recherche experimentale sur les composantes assurant la specificite du chant chez *Phylloscopus bonelli. Abstr. XV Vongr. Int. Ornith.* 73.

Broekhuysen, G. J., 1966. The avifauna of the Cape "Protea-Macchia" heath habitat in South Africa. *Ostr. Suppl.* #6:323–334.

Brown, W. L., and E. O. Wilson, 1956. Character displacement. *Syst. Zool.* 5:49–65.

Buffon, Compte de, 1812. *Histoire Naturelle XI.* History of Birds, Vol. Transl. W. Smellie. Cadell and Davies. London.

Burt, W. H., and R. P. Grossenheider, 1952. *A Field Guide to the Mammals.* Boston, Houghton-Mifflin.

Cade, T. J., 1960. Ecology of the Peregrine Falcon and Gyr Falcon populations in Alaska. *Univ. Calif. Publ. Zoology* 63:151–290.

Chapin, J. P., 1932. *The Birds of the Belgian Congo. Pt. I.* Bull. Amer. Mus. Nat. Hist. LXV:1–756.

Chapman, F. M., 1923. Mutation among birds of the genus *Buarremon. Bull. Amer. Mus. Nat. Hist.* 48:243–278.

Cheke, R. A., 1971. Feeding ecology and significance of interspecific territoriality of African montane sunbirds (Nectariniidae). *Rev. Zool. Bot. Afr.* 84:50–64.

Cody, M. L., 1966. The consistency of inter- and intra-specific continental bird species counts. *Amer. Natur.* 100:371–376.

Cody, M. L., 1968a. On the methods of resource division in grassland bird communities. *Amer. Natur.* 102:107–147.

Cody, M. L., 1968b. Interspecific territoriality among hummingbird species. *Condor* 70:270–271.

REFERENCES

Cody, M. L., 1969. Convergent characteristics in sympatric populations: A possible relation to interspecific territoriality. *Condor* 71:222–239.

Cody, M. L., 1970. Chilean bird distributions. *Ecology* 51:455–464.

Cody, M. L., 1971. Finch flocks in the Mohave Desert. *Theor. Pop. Biol.* 2:142–158.

Cody, M. L., 1973a. Coexistence, coevolution and convergent evolution in seabird communities. *Ecology* 54:31–44.

Cody, M. L., 1973b. Optimization in ecology. *Science;* in press.

Cody, M. L., and C. B. J. Cody, 1972. Territory size, food density and clutch size in island wren populations. *Condor* 75:473–477.

Cody, M. L., and J. H. Brown, 1969. Song asynchrony in chaparral birds. *Nature* 222:778–780.

Cody, M. L., and J. H. Brown, 1970. Character convergence in Mexican finches. *Evolution* 24:304–310.

Colwell, R. K., and F. J. Futuyma, 1971. On the measurement of niche breadth and overlap. *Ecology* 52:567–576.

Craighead, J. J., and F. C. Craighead, 1956. *Hawks, Owls and Wildlife*. Philadelphia, Stackpole.

Crowell, K., 1962. Reduced interspecific competition among the birds of Bermuda. *Ecology* 43:75–88.

Darlington, P. J., 1965. *The Biogeography of the Southern End of the World*. Cambridge, Mass., Harvard.

Darwin, C. R., 1859. *On the Origin of Species*. London.

Dethier, V. G., 1954. Evolution of feeding preferences in phytophagous insects. *Evolution* 8:33–54.

Diamond, J. M., 1970a. Ecological consequences of island colonization by southwest Pacific birds. I. Types of niche shifts. *Nat. Acad. Sci. Proc.* 67:529–536.

Diamond, J. M., 1970b. Ecological consequences of island colonization by southwest Pacific birds. II. The effect of species diversity on total population density. *Nat. Acad. Sci. Proc.* 67:1715–1721.

REFERENCES

Diamond, J. M., 1972. The avifauna of the eastern highlands of New Guinea. *Publ. Nuttall Ornith. Club,* Cambridge, Mass.

Diamond, J. M., and J. Terborgh, 1967. Observations on bird distribution and feeding assemblages along the Rio Callaria, Department of Loreto, Peru. *Wilson Bull.* 79:273–282.

DiCastri, F., and H. Mooney, 1973. *Ecological Studies #7.* Berlin, Springer Verlag.

Drury, W. H. Jr., 1961. Studies on the breeding biology of the horned lark, water pipit, lapland longspur and snow bunting on Bylot Island, Northwest Territories, Canada. *Bird Banding* 32:1–46.

Elton, C., 1927. *Animal Ecology,* London, Sidgewick and Jackson.

Emlen, J. M., 1966. The role of time and energy in food preference. *Amer. Natur.* 100:611–617.

Emlen, J. M., 1968. Optimal choice in animals. *Amer. Natur.* 102:385–389.

Emlen, J. T., 1972. Size and structure of a wintering avian community in southern Texas. *Ecology* 53:317–329.

Emlen, S. T., 1971. The role of song in individual recognition in the indigo bunting. *Z. Tierpsychol.* 28:241–246.

Ferry, C., and A. Deschaintre, 1966. *Hippolais icterina* et *polyglotta* dans leur zone de sympatrie. *Abstr. XIV Int. Congr. Ornith.* 57–58.

Filippi, G., 1847. *Mus. Mediol., Anim. Vertebr., cl.2, Aves,* pp. 15, 32. (cited in Johnson, A., *et al.,* 1957).

Fretwell, S., 1972. *Populations in a Seasonal Environment.* Monogr. Pop. Biol., Princeton Univ. Press.

Friedmann, H., 1946. Ecological counterparts in birds. *Sci. Monthly* 43:395–398.

Fry, C. H., 1970. Convergence between jacamars and bee-eaters. *Ibis* 112:257–259.

Gause, G. F., 1934. *The Struggle for Existence.* New York, Haffner.

REFERENCES

Gibb, J., 1954. Feeding ecology of tits, with notes on the tree-creeper and goldcrest. *Ibis* 96:513–543.

Goodall, D. W., 1970. Statistical plant ecology. *Ann. Rev. Ecol. Syst.* 1:99–124.

Grant, P. R., 1966. The coexistence of two wren species of the genus *Thryothorus. Wilson Bull.* 78:266–278.

Grant, P. R., 1968. Polyhedral territories of animals. *Amer. Natur.* 102:75–80.

Grant, P. R., 1972. Convergent and divergent character displacement. *Biol. J. Linn. Soc.* 4:39–68.

Grinnell, J., 1904. The origin and distribution of the chestnut-backed chickadee. *Auk* 21:364–382.

Grinnell, J., 1917. The niche-relationships of the California thrasher. *Auk* 34:427–433.

Gysels, H., and M. Rabaey, 1964. Taxonomic relationships of *Alca torda, Fratercula arctica* and *Uria aalge* as revealed by biochemical methods. *Ibis* 106:536–540.

Hairston, N., F. Smith, and L. Slobodkin, 1960. Community structure, population control and competition. *Amer. Natur.* 94:421–425.

Hall, B., R. E. Moreau, and J. Galbraith, 1966. Polymorphism and parallelism in the African bushshrikes of the genus *Malaconotus* (including *Chlorophoneus*). *Ibis* 108:161–182.

Harris, R. D., 1944. The chestnut-collared longspur in Manitoba. *Wilson Bull.* 56:105–115.

Hespenheide, H., 1966. The selection of seed size by finches. *Wilson Bull.* 78:191–197.

Hespenheide, H., 1971. Food preference and the extent of overlap in some insectivorous birds with special reference to the Tyranidae. *Ibis* 113:59–72.

Holmes, R. T., and F. A. Pitelka, 1968. Food overlap among coexisting sandpipers on northern Alaskan tundra. *Syst. Zool.* 17:305–318.

Horn, H. S., 1968. The adaptive significance of colonial nesting in the Brewer's Blackbird (*Euphagus cyanocephalus*). *Ecology* 49:682–694.

Hutchinson, G. E., 1941. Ecological aspects of succession in natural populations. *Amer. Natur.* 75:406–418.

Hutchinson, G. E., 1958. Concluding remarks. *Cold Spring Harbor Symp. Quant. Biol.* 22:415–427.

Hutchinson, G. E., 1959. Homage to Santa Rosalia, or Why are there so many kinds of animals? *Amer. Natur.* 93:145–159.

Hutchinson, G. E., 1961. The paradox of the plankton. *Amer. Natur.* 95:137–145.

Ingham, C. O., 1963. An ecological and taxonomic study of the ants of the Great Basin and Mohave Desert regions of southwestern Utah. *Dissert. Abstr.* 24:1759–1760.

Ingolfsson, A., 1970. The feeding ecology of large gulls (*Larus*) in Iceland. *Abstr. Int. Congr. Ornith. XV*:124.

Joensen, A. H., 1965. (An investigation on bird populations in four deciduous forest areas on Als in 1962 and 1963). *Dansk Ornith. Foren. Tidsskr.* 59:111–186.

Johnson, A., J. D. Goodall, and R. A. Philippi, 1957. *Las Aves de Chile.* Platt Establ., Buenos Aires.

Karr, J. R., 1972. A comparative study of the structure of avian communities in selected Panamanian and Illinois habitats. *Ecol. Monogr.* 41:207–233.

Kear, J., 1962. Food selection in finches with special reference to interspecific differences. *Proc. Zool. Soc. London* 138:163–204.

Kendeigh, S. C., 1947. Bird population studies in the coniferous forest biome during a spruce budworm outbreak. *Biol. Bull.* 1:1–100.

Kikkawa, J., 1966. Population distribution of land birds in temperate rain forest of southern New Zealand. *Trans. Royal Soc. New Zealand* 7:215–277.

Kikkawa, J., I. Hore-Lacy, and J. LeGay Brereton, 1965. A

preliminary report on the birds of the New England National Park. *Emu* 65:139–143.

Klauber, L. M., 1972. *Rattlesnakes.* Vol. 1, 2nd Ed. Berkeley, Univ. Calif. Press.

Klopfer, P. H., and R. H. MacArthur, 1961. On the causes of tropical species diversity: Niche overlap. *Amer. Natur.* 95:223–226.

Kruuk, H., 1967. Competition for food between vultures in East Africa. *Ardea* 55:171–203.

Kuroda, N. h., 1967. Morpho-anatomical analysis of parallel evolution between Diving Petrel and Ancient Auk, with comparative osteological data of other species. *Misc. Rep. Yamashina Inst. Ornith. Zool.* 5,2(28):111–137.

Kuroda, N. h., 1968. Avifaunal survey of Mt. Iwate, northern Honshu. *Misc. Rep. Yamashina Inst. Ornith. Zool.* 5,3(29): 214–240.

Kusnetsov, A. P., 1970. Ecology and distribution of the sea bottom fauna and flora. *Trans. Shirshov Inst. Ocean.* 88:1–112.

Kusnezov, N., 1953. Formas de vida especializadas y su desarollo en differentes partes del mundo. *Dusenia* (Brasil) 4(2):85–102.

Kusnezov, N., 1956. A comparative study of ants in desert regions of central Asia and of South America. *Amer. Natur.* 90:349–360.

Lack, D. L., 1934. Habitat distribution in certain Icelandic birds. *J. Anim. Ecol.* 3:81–90.

Lack, D. L., 1946. Competition for food in birds of prey. *J. Anim. Ecol.* 15:123–129.

Lack, D. L., 1947. *Darwin's Finches.* Cambridge. Univ. Press.

Lack, D. L., 1966. *Population Studies of Birds.* Oxford. Univ. Press.

Lack, D. L., 1968. *Ecological Aspects of Reproduction in Birds.* London, Methuen.

REFERENCES

Lack, D. L., 1969. Tit niches in two worlds, or, Homage to G. E. Hutchinson. *Amer. Natur.* 103:43–49.

Lack, D. L., 1971. *Ecological Isolation in Birds.* Oxford, Blackwell's.

Land, H. C., 1963. A tropical feeding tree. *Wilson Bull.* 75:199–200.

Lanyon, W. E., 1957. The comparative biology of the meadowlarks (*Sturnella*) in Wisconsin. *Nuttall Ornith. Club Publ.* 1:1–67.

Lamprey, H. F., 1963. Ecological separation of the large mammal species in the Taragire Game Preserve, Tanganyika. *East Afr. Wildl. J.* 1:63–92.

Lein, M. R., 1972. A trophic comparison of avifaunas. *Syst. Zool.* 21:135–150.

Lemon, R. W., and A. Herzog, 1969. The vocal behavior of cardinals and pyrrhuloxias in Texas. *Condor* 71:1–15.

Levins, R., 1968. *Evolution in Changing Environments.* Monogr. Pop. Biol., Princeton Univ. Press.

Levins, R., and R. H. MacArthur, 1969. An hypothesis to explain the incidence of monophagy. *Ecology* 50:910–911.

Leyhausen, P., and R. Wolff, 1959. Das Revier einer Hauskatz. *Z. Tierpsychol.* 16:66–70.

Linnaeus, C., 1758. *Systema Natura. Regnum Animale.* 10th. Ed. Leipsig, Engelmann.

Lorenz, K., 1966. *On Aggression.* New York, Bantam.

Low, R. M., 1971. Interspecific territoriality in a pomacentrid reef fish, *Pomacentrus flavicauda* Whitley. *Ecology* 52:648–654.

Luce, R., 1959. On the possible psychophysical laws. *Psych. Rev.* 66:81–95.

MacArthur, R. H., 1958. Population ecology of some warblers of northeastern coniferous forests. *Ecology* 39:599–619.

MacArthur, R. H., 1960. On the relative abundance of species. *Amer. Natur.* 94:25–36.

MacArthur, R. H., 1965. Patterns in species diversity. *Biol. Rev.* 40:510–533.

REFERENCES

MacArthur, R. H., 1968. The theory of the niche. *In Population Biology and Evolution.* R. Lewontin, Ed. New York, Syracuse Univ. Press.

MacArthur, R. H., 1970. Species packing and competitive equilibria for many species. *Theor. Pop. Biol.* 1:1–11.

MacArthur, R. H., 1972. *Geographic Ecology.* New York, Harper and Row.

MacArthur, R. H., J. M. Diamond, and J. Karr, 1972. Density compensation in island faunas. *Ecology* 53:330–342.

MacArthur, R. H., and R. Levins, 1967. The limiting similarity, convergence and divergence of coexisting species. *Amer. Natur.* 101:377–385.

MacArthur, R. H., and J. MacArthur, 1961. On bird species diversity. *Ecology* 42:594–598.

MacArthur, R. H., and E. R. Pianka, 1966. On optimal use of a patchy environment. *Amer. Natur.* 100:603–607.

MacArthur, R. H., H. Recher, and M. L. Cody, 1966. On the relation between habitat selection and bird species diversity. *Amer. Natur.* 100:319–332.

MacNaughton, S. J., and L. L. Wolf, 1970. Dominance and the niche in ecological systems. *Science* 167:131–139.

Marler, P. R., 1960. In: *Animal Sounds and Communication.* W. E. Lanyon and W. N. Tavolga, Eds. A.I.B.S., Washington.

Marshall, J. T. Jr., 1960. Interrelations of Abert and brown towhees. *Condor* 62:49–64.

May, R. M., and R. H. MacArthur, 1972. Niche overlap as a function of environmental variability. *Nat. Acad. Sci. Proc.* 69:1109–1113.

Mooney, H. A., and E. L. Dunn, 1970. Convergent evolution of Mediterranean-climate evergreen sclerophyll shrubs. *Evolution* 24:292–303.

Moreau, R. E., 1948. Ecological isolation in a rich tropical avifauna. *J. Anim. Ecol.* 17:113–126.

Moreau, R. E., 1966. *The Bird Faunas of Africa and Its Islands.* London, Academic Press.

REFERENCES

Moynihan, M., 1960. Some adaptations which help to promote gregariousness. *Proc. XII Int. Congr. Ornith.* 523–541.

Moynihan, M., 1962. The organization and probable evolution of some mixed species flocks of neotropical birds. *Smithsonian Misc. Coll.* 143:1–140.

Moynihan, M., 1963. Interspecific relations between some Andean birds. *Ibis* 105:327–339.

Moynihan, M., 1968. Social mimicry: Character convergence versus character displacement. *Evolution* 22:315–331.

Murphy, R. C., 1936. *The Oceanic Birds of South America.* New York, Amer. Mus. Nat. Hist.

Murray, B., 1971. The ecological consequences of interspecific territorial behavior in birds. *Ecology* 52:414–423.

Nakamura, T., 1963. A survey of an upland grassland bird community during the breeding season. *Misc. Rep. Yamashina Inst. Ornith. Zool.* 3,5(20):334–357.

Nakamura, T., S. Yamaguchi, K., Iijima, and T. Kagawa, 1968. A comparative study on the habitat preference and home range of four species of *Emberiza* on peat grassland. *Misc. Rep. Yamashina Inst. Ornith. Zool.* 5,4(30):313–336.

Naumann, J. F., 1903. *Naturgeschichte der Vögel Mitteleuropas.* vol. 9. Leipzig.

Newton, I., 1967. The adaptive radiation and feeding ecology of the British finches. *Ibis* 109:33–98.

Nice, N. M., 1943. *Studies in the Life History of the Song Sparrow* Vol. II. Trans. Linn. Soc. New York.

Nicholson, E. M., 1932. *The Art of Bird Watching.*

Nørrevang, A., 1960. (Habitat selection of seabirds in Mykines, Faeroes). *Dansk Ornith. Foren. Tidsskr.* 54:9–35.

Ogasawara, K., 1966. Bird survey of Mt. Kurikoma and its surrounding area, northern Honshu, with ecological notes. *Misc. Rep. Yamashina Inst. Ornith. Zool.* 4,5(25):371–377.

Orians, G. H., 1969. On the evolution of mating systems in birds and mammals. *Amer. Natur.* 103:589–603.

Orians, G. H., and H. Horn, 1969. Overlap in foods and

foraging of four species of blackbirds in the Potholes of central Washington. *Ecology* 50:930–938.

Orians, G. H., and M. F. Willson, 1964. Interspecific territories of birds. *Ecology* 45:736–745.

Paine, R. T., 1966. Food web complexity and species diversity. *Amer. Natur.* 100:65–75.

Patrick, R., 1961. A study of the numbers and kinds of species found in rivers in eastern United States. *Proc. Acad. Nat. Sci. Philadelphia* 113(10):215–258.

Patrick, R., 1964. A discussion of the results of the Catherwood expedition to the Peruvian headwaters of the Amazon. *Proc. Internat. Assoc. Theor. Appl. Limnol. XV*(2):1084–1090.

Perrins, C. M., 1970. The timing of bird's breeding seasons. *Ibis* 112:242–255.

Peters, J. L., 1968. *Checklist of Birds of the World, XIV.* Worcester, Mass., Heffernan Press.

Peterson, R. T., 1961. *A Field Guide to the Birds of the Western United States.* Boston, Houghton-Mifflin.

Pianka, E. R., 1966. Convexity, desert lizards, and spatial heterogeneity. *Ecology* 47:1055–1059.

Pianka, E. R., 1967. On lizard species diversity: North American flatland deserts. *Ecology* 48:333–351.

Pianka, E. R., 1969a. Habitat specificity, speciation and species diversity in Australian desert lizards. *Ecology* 50:498–502.

Pianka, E. R., 1969b. Sympatry of desert lizards (*Ctenotus*) in Western Australia. *Ecology* 50:1012–1030.

Pianka, E. R., 1970. The ecology of *Moloch horridus* (Lacertilia:Agamidae) in Western Australia. *Copeia* 1970(1):90–103.

Pianka, E. R., 1971. Comparative ecology of two lizards. *Copeia* 1971(1):129–138.

Pico, M. M., D. Maldonado, and R. Levins, 1965. Ecology and genetics of Puerto Rican *Drosophila:* I. Food preference of sympatric species. *Carib. J. Science* 5:29–37.

REFERENCES

Pitleka, F. A., P. Q. Tomich, and G. W. Triechel, 1955. Ecological relations of jaegers and owls as lemming predators near Barrow, Alaska. *Ecol. Monogr.* 25:85–117.

Pulliam, H. R., and F. Enders, 1971. The feeding ecology of five sympatric finch species. *Ecology* 52:557–566.

Rasa, O. A. E., 1969. Territoriality and the establishment of dominance by means of visual cues in *Pomacentrus jenkinsi* (Pisces; Pomacentridae) *Z. Tierpsychol.* 26:825–845.

Recher, H., 1969. Bird species diversity and habitat diversity in Australia and North America. *Amer. Natur.* 103:75–79.

Richdale, L. E., 1943. The kuaka or diving petrel, *Pelecanoides urinatrix* (Gmelin). *Emu* 43:24–48.

Ricklefs, R. E., 1966. The temporal component of diversity among species of birds. *Evolution* 20:235–242.

Ridgeway, R., 1901. *The Birds of North and Middle America.* Bull. U.S. National Mus. #50, Pt. 1.

Ridpath, M. G., and R. E. Moreau, 1966. The birds of Tasmania: Ecology and evolution. *Ibis* 108:348–393.

Roughgarden, J., 1973. On invading a guild of coexisting species. *Theor. Pop. Biol.* (in press).

Rosenzweig, M. R., 1968a. Anecdotal evidence for the reality of character convergence. *Amer. Natur.* 102:491–492.

Rosenzweig, M. R., 1968b. Net primary productivity of terrestrial communities: Prediction from climatological data. *Amer. Natur.* 102:67–74.

Sage, R., 1971. Preliminary considerations on the ecological convergence of the lizard faunas of the chaparral communities in Chile and California. MS.

Schmidt, K. P., 1922. *The Amphibians and Reptiles of Lower California.* Bull. U.S. National Mus. 46:607–707.

Schoener, T., 1965. The evolution of bill size differences among sympatric species of birds. *Evolution* 19:189–213.

Schoener, T., 1968. The *Anolis* lizards of Bimini: Resource partitioning in a complex fauna. *Ecology* 49:704–726.

Schoener, T., 1970. Size patterns in West Indian *Anolis*

lizards. II. Correlations with the sizes of particular sympatric species—displacement and convergence. *Amer. Natur.* 104:155–174.

Schoener, T., 1971. On the theory of feeding strategies. *Ann. Rev. Ecol. Syst.* 2:369–404.

Schoener, T., and G. C. Gorman, 1968. Some niche differences in three lesser Antillean lizards of the genus *Anolis. Ecology* 49:819–830.

Selander, R., 1966. Sexual dimorphism and differential niche utilization in birds. *Condor* 68:113–151.

Sergeant, D. E., 1951. Ecological relationships of the guillemots *Uria aalge* and *Uria lomvia. Proc. X Int. Congr. Ornith.* 578–587.

Sheldon, A. L., 1968. Species diversity and longitudinal succession in stream fishes. *Ecology* 49:193–198.

Short, L., 1968. Sympatry of red-breasted meadowlarks in Argentina, and the taxonomy of meadowlarks (Aves: *Leistes, Pezites* and *Sturnella*). *Amer. Mus. Nov.* 2349:1–30.

Sibley, C. G., 1950. Species formation in the red-eyed towhees of Mexico. *Univ. Calif. Publ. Zool.* 50:109–194.

Slobodkin, L. R., and H. Sanders, 1969. On the contribution of environmental predictability to species diversity. *In Diversity and Stability in Ecological Systems.* Brookhaven Symp. Biology 22.

Smith, W. J., and F. Vuilleumier, 1971. Evolutionary relationships of some South American ground tyrants. *Bull. Mus. Comp. Zool.* 141:179–268.

Soikkeli, M., 1965. On the structure of the bird fauna on some coastal meadows in western Finland. *Orn. Fenn.* 42:101–111.

Sokal, R., and P. Sneath, 1963. *Principles of Numerical Taxonomy.* San Francisco, Freeman.

Southern, H. N., and I. Linn, 1964. Distribution, range, habitat. *In Handbook of British Mammals.* M. Southern, Ed. Blackwell's, Oxford.

Steere, J. B., 1894. On the distribution of genera and species of non-migratory land-birds in the Philippines. *Ibis* 1894:411–420.

Stiles, F. G., 1973. Food supply and the annual cycle of the Anna hummingbird. *Univ. Calif. Publ. Zool.* 97:1–109.

Storer, R. W., 1966. Sexual dimorphism and food habits in three North American accipiters. *Auk* 83:423–436.

Stresemann, E., 1950. Interspecific competition in chats. *Ibis* 92:148.

Terborgh, J., 1971. Distribution on environmental gradients: Theory and a preliminary interpretation of distributional patterns in the Avifauna of the Cordillera Vilcabamba, Peru. *Ecology* 52:23–40.

Terborgh, J., and J. M. Diamond, 1970. Niche overlap in feeding assemblages of New Guinea birds. *Wilson Bull.* 82:29–52.

Thönen, W., 1962. Studien über die Mönchsmeise. Ornith. Beobach. 59:103–172.

Thorson, G., 1957. Bottom communities (sublittoral or shallow shelf). *In Treatise on Marine Ecology and Palaeoecology.* J. W. Hedgpeth, Ed. Ch. 17. *Mem. Geol. Soc. Amer.* 67:461–534.

Thorson, G., 1958, 1960. Parallel level bottom communities, their temperature adaptation, and the balance between predators and food animals. *Persp. Marine Biology,* Berkeley.

Turc, L., 1955. Lebilan d'eau des sols. *Ann. Agron.* 6:5. *Rep. Lamashina Inst. Ornith. Zool.* 3,1(16):1–32.

Uramoto, M., 1961. Ecological study of the bird community of the broadleaved decoduous forest of central Japan. *Misc. Rep. Yamashina Inst. Ornith. Zool.* 3,1(16):1–32.

Vandermeer, J., 1972. The covariance of the community matrix. *Ecology* 53:187–189.

Vesey-Fitzgerald, D. F., 1965. The utilization of natural pastures by wild animals in the Rukwa Valley, Tanganyika. *East Afr. Wildl. J.* 3:38–48.

Vuilleumier, F., 1972. Bird species diversity in Patagonia (temperate South America). *Amer. Natur.* 106:266–271.

REFERENCES

Walker, E., 1964. *The Mammals of the World.* Vols. I–III. Baltimore, Johns Hopkins.

Wallace, A. R., 1869. *The Malay Archipelago.* New York, Harper.

Wallgren, M., 1954. Energy metabolism of two species of the genus *Emberiza* as correlated with distribution and migration. *Acta Zool. Fenn.* 84:1–110.

Wetmore, A., 1943. The birds of southern Veracruz, Mexico. *Proc. U.S. Mus. Nat. Hist.* 93:215–340.

White, C. M. N., 1951. Weaver birds at Lake Mweru. *Ibis* 93:626–627.

Wiens, J., 1969. An approach to the study of ecological relationships among grassland birds. *Ornith. Monogr.* 8, Amer. Ornith. Union.

Williams, A. B., 1936. The composition and dynamics of a beech-maple climax community. *Ecol. Monogr.* 6:318–408.

Williams, F., 1971. *Systems Analysis and Simulation in Ecology.* B. Patten, Ed. New York, Academic Press, pp. 197–267.

Williamson, K., 1967. A bird community of accreting sand dunes and salt marsh. *Brit. Birds* 60:145–157.

Willis, E. O., 1966. Competitive exclusion and birds at fruiting trees in western Colombia. *Auk* 83:479–480.

Willson, M. F., 1971. Seed selection in some North American finches. *Condor* 73:415–429.

Winterbottom, J. M., 1966. Ecological distribution of birds in the indiginous vegetation of the southwest Cape. *Ostrich* 37:76–91.

Wynne-Edwards, V. C., 1962. *Animal Dispersion in Relation to Social Behavior.* Edinburgh, Oliver and Boyd.

Yeaton, R. I., 1972. An ecological analysis of chaparral and pine forest bird communities on Santa Cruz Island and mainland California. Ph. D. Thesis, University of California at Los Angeles.

Yeaton, R. I., and M. L. Cody, 1973. Competitive release in island song sparrow populations. *Theor. Pop. Biol.* (in press).

Author Index

Albrecht, H., 258
Alm, B., 177
Arnold, S., 259
Ashmole, N. P., 36
Asplund, K., 259

Baird, S. F., 49
Bedard, J., 208
Belopolsk'ii, L. O., 205, 208
Bernard, F., 172
Bernstein, R. A., 172
Blondel, J., 201
Bowman, R. I., 35
Bremond, J. C., 247
Broekhuysen, G. J., 201
Brown, J. H., 48, 221, 224, 243, 245
Brown, W. L., 6, 78
Buffon, Compte de, 166
Bundick, B., 17
Burt, W. H., 133

Cade, T. J., 218
Catesby, 166
Chapin, J. P., 203
Chapman, F. M., 243
Cheke, R. A., 245
Cody, C. B. J., 141
Colwell, R. K., 71, 72
Craighead, J. J., 217
Crowell, K., 132, 150

Darlington, P. J., 180
Darwin, C. R., 49, 213
Dethier, V. G., 56
Diamond, J. M., 16, 17, 38, 65, 132, 140, 181, 203
DiCastri, F., 189
Drury, W. H., 217, 218

Elton, C., 50
Emlen, J. M., 56
Emlen, J. T., 158
Emlen, S. T., 249
Enders, F., 68

Ferry, C., 247
Filippi, G., 166
Fretwell, S., 159, 261
Friedmann, H., 166
Fry, C. H., 166, 202

Gause, G. F., 51, 54, 210
Gibb, J., 30
Goodall, D. W., 92
Grant, P. R., 130, 150, 217, 246
Grinnell, J., 50, 262
Gysels, H., 175

Hairston, N., 213
Hall, B., 241
Harper, K., 214
Harris, R. D., 217, 218
Hespenheide, H., 36, 37, 38
Holmes, R. T., 38, 39
Horn, H. S., 120, 204
Hutchinson, G. E., 51, 210, 255

Ingham, C. O., 172
Ingolfsson, A., 207

Jenkins, R., 257
Joensen, A. H., 180, 185
Johnson, A., 169

Karr, J. R., 30
Kear, J., 36
Kendeigh, S. C., 210
Kikkawa, J., 181, 185
Klauber, L. M., 136
Klopfer, P. H., 255
Kruuk, H., 204, 207
Kuroda, N. h., 173, 177
Kusnetsov, A. P., 171
Kusnezov, N., 171

Lack, D. L., 7, 35, 38, 47, 132, 166, 168, 202, 207, 208, 248, 260
Lamprey, H. F., 213
Land, H. C., 203

305

Subject Index

Bird Index

310

Library of Congress Cataloging in Publication Data

Cody, Martin L   1941–
    Competition and the structure of bird communities.

    (Monographs in population biology, no. 7)
    Bibliography: p.
    1. Bird populations.   2. Birds—Behavior.
I. Title.   II. Series.
QL677.4.C63   598.2′5′24   73-16202
ISBN 0-691-08134-4
ISBN 0-691-08135-2 (pbk.)